SpringerBriefs in Electrical and Computer Engineering

SpringerBriefs in Speech Technology

Series Editor:
Amy Neustein

For further volumes:
http://www.springer.com/series/10059

Editor's Note

The authors of this series have been hand-selected. They comprise some of the most outstanding scientists – drawn from academia and private industry – whose research is marked by its novelty, applicability, and practicality in providing broad based speech solutions. The SpringerBriefs in Speech Technology series provides the latest findings in speech technology gleaned from comprehensive literature reviews and *empirical investigations* that are performed in both laboratory and *real life* settings. Some of the topics covered in this series include the presentation of real life commercial deployment of spoken dialog systems, contemporary methods of speech parameterization, developments in information security for automated speech, forensic speaker recognition, use of sophisticated speech analytics in call centers, and an exploration of new methods of soft computing for improving human-computer interaction. Those in academia, the private sector, the self service industry, law enforcement, and government intelligence, are among the principal audience for this series, which is designed to serve as an important and essential reference guide for speech developers, system designers, speech engineers, linguists and others. In particular, a major audience of readers will consist of researchers and technical experts in the automated call center industry where speech processing is a key component to the functioning of customer care contact centers.

Amy Neustein, Ph.D., serves as Editor-in-Chief of the International Journal of Speech Technology (Springer). She edited the recently published book "Advances in Speech Recognition: Mobile Environments, Call Centers and Clinics" (Springer 2010), and serves as guest columnist on speech processing for Womensenews. Dr. Neustein is Founder and CEO of Linguistic Technology Systems, a NJ-based think tank for intelligent design of advanced natural language based emotion-detection software to improve human response in monitoring recorded conversations of terror suspects and helpline calls. Dr. Neustein's work appears in the peer review literature and in industry and mass media publications. Her academic books, which cover a range of political, social and legal topics, have been cited in the Chronicles of Higher Education, and have won her a pro Humanitate Literary Award. She serves on the visiting faculty of the National Judicial College and as a plenary speaker at conferences in artificial intelligence and computing. Dr. Neustein is a member of MIR (machine intelligence research) Labs, which does advanced work in computer technology to assist underdeveloped countries in improving their ability to cope with famine, disease/illness, and political and social affliction. She is a founding member of the New York City Speech Processing Consortium, a newly formed group of NY-based companies, publishing houses, and researchers dedicated to advancing speech technology research and development.

Todor Ganchev

Contemporary Methods
for Speech Parameterization

 Springer

Todor Ganchev
Wire Communications Laboratory
Department of Electrical & Computer Engineering
University of Patras
Rion-Patras, Greece
tganchev@ieee.org

ISSN 2191-8112   e-ISSN 2191-8120
ISBN 978-1-4419-8446-3   e-ISBN 978-1-4419-8447-0
DOI 10.1007/978-1-4419-8447-0
Springer New York Dordrecht Heidelberg London

Library of Congress Control Number: 2011929873

Printed on acid-free paper

Springer is part of Springer Science+Business Media (www.springer.com)

Preface

The preparation of the present brief book was motivated by the significant and long-standing interest of the speech processing community to short-time cepstrum-based parameterization of speech. In approximately 100 pages, this volume brings together relevant information about 11 speech parameterization techniques and some of their variants that emerged during the past decades. The unification of presentation style and notations, along with some additional details, numerous illustrations, and comparative descriptions, are expected to facilitate the understanding of the properties and the range of applicability of these speech parameterization methods.

With the intention to provide a self-contained description of each individual speech parameterization method, the author took the liberty to reiterate the explanation of the processing steps which are common among numerous speech features. In this way, the reader will see repeatedly similar comments and explanations about the discrete Fourier transform, the discrete wavelet packet transform, the discrete cosine transform, etc. This redundancy was introduced with the idea to facilitate the reader who does not want to browse through half of the book in order to comprehend how exactly certain speech features are computed. In this manner, the repetitions in the exposition of certain processing steps shall be seen as a way to avoid the frequent forwarding to the explanation of other speech parameterizations. The author hopes this presentation style will not irritate the meticulous reader who will read the entire book.

Besides the comprehensive exposition of the 11 speech parameterizations of interest, this volume also offers a comparative analysis of the similarities and differences in their computation and performance evaluation in common experimental conditions on three dissimilar speech processing tasks. The empirical comparison of these speech features in three different setups intends to enhance the standpoint of the large community of speech processing researchers and practitioners working in the field of speech and speaker recognition or other related speech processing tasks.

Although this volume is not aimed or organized as a university textbook, it could be used as a comprehensive source of information by faculty and teaching staff involved in undergraduate or postgraduate courses that embrace topics on Natural Language Processing or Speech Technology. Postgraduate students may wish to use this book for deepening their knowledge and understanding on the various speech parameterization techniques. Engineers and developers of speech processing applications may gain better understanding of the motivation behind the widely used speech parameterizations and their range of applicability. The author hopes that the interested reader will find some less-known facts about the origins of speech parameterization and will enjoy the few clarifications about the creative contributors whose original ideas sometimes remain unacknowledged.

The readers who notice errors or inaccuracies in the exposition and report these back to me will receive a postcard from Greece with acknowledgment for their contribution and their names will appear in the acknowledgment section of the subsequent extended editions of this book. The more energetic critics of this work will also receive a tiny souvenir from Greece.

In conclusion, I would like to acknowledge that this book would not exist without the support of my spouse, Tsenka Stoyanova, during the long evenings and many weekends I spent on this project. Lastly, I would like to acknowledge the contribution of my long-time collaborators: Dr. Mihalis Siafarikas, who prepared the source code for four out of the five wavelet packet transform-based speech parameterizations considered in this book, and Dr. Iosif Mporas, who performed the speech recognition evaluation reported in Sect. 6.

Patras, Greece   Todor Ganchev
February, 2011

Contents

Acronyms

ACE	Advanced combinational encoder
CMS	Cepstral mean subtraction
DCFopt	Optimal decision cost
DFT	Discrete Fourier transform
DRN	Dynamic range normalization
DWPT	Discrete wavelet packet transform
EER	Equal error rate
ERB	Equivalent rectangular bandwidth
FA	False acceptance
FB	Filter-bank
FR	False rejection
GMM	Gaussian mixture model
HFCC	Human factor cepstral coefficients
HMM	Hidden Markov models
HTK	Cambridge HMM toolkit
LFCC	Linear-frequency cepstral coefficients
LPC	Linear prediction coefficients
LPCC	Linear predictive cepstral coefficients
MAPSSWE	Matched pairs sentence segment word error test
MFCC	Mel frequency cepstral coefficients
NIST	The National Institute of Standards and Technology of the USA
ODWPT	Overlapping discrete wavelet packet transform
PLP	Perceptual linear prediction
RCC	Real cepstral coefficients

SBC   Sub-band cepstral parameters
SNR   Signal-to-noise ratio
SRE   Speaker recognition evaluation
stDFT   Short-time discrete Fourier transform

WPF   Wavelet packet transform-based speech features
WPF-ACE   Wavelet packet transform-based speech features employing the
             filter-bank of Nogueira et al. (2006)
WPF-FD   Wavelet packet transform-based speech features of Farooq-Datta
WPF-OBJ   Wavelet packet transform-based objectively optimized speech
             features
WPF-OVL   Wavelet packet transform-based speech features with overlapping
             sub-bands
WPF-SBC   Wavelet packet transform-based sub-band cepstral parameters

Contemporary Methods for Speech Parameterization

Short-Time Cepstrum-Based Speech Features

Abstract This brief book offers a general view of short-time cepstrum-based speech parameterization and provides a common ground for further in-depth studies on the subject. Specifically, it offers a comprehensive description, comparative analysis, and empirical performance evaluation of eleven contemporary speech parameterization methods, which compute short-time cepstrum-based speech features. Among these are five discrete wavelet packet transform (DWPT)-based and six discrete Fourier transform (DFT)-based speech features and some of their variants which have been used on the speech recognition, speaker recognition, and other related speech processing tasks. The main similarities and differences in their computation are discussed and empirical results from performance evaluation in common experimental conditions are presented. The recognition accuracy obtained on the monophone recognition, continuous speech recognition, and speaker recognition tasks is contrasted against the one obtained for the well-known and widely used Mel Frequency Cepstral Coefficients (MFCC). It is shown that many of these methods lead to speech features that do offer competitive performance on a certain speech processing setup when compared to the venerable MFCC. The last does not target the promotion of certain speech features but instead aims to enhance the common understanding about the advantages and disadvantages of the various speech parameterization techniques available today and to provide the basis for selection of an appropriate speech parameterization in each particular case.

In brief, this volume consists of nine sections. Section 1 summarizes the main concepts on which the contemporary speech parameterization is based and offers some background information about their origins. Section 2 introduces the objectives of speech pre-processing and describes the processing steps that are commonly used in the contemporary speech parameterization methods. Sections 3 and 4 offer a comprehensive description and a comparative analysis of the DFT- and DWPT-based speech parameterization methods of interest. Sections 5–7, present results from experimental evaluation on the monophone recognition, continuous speech recognition, and speaker recognition tasks, respectively. Section 8 offers concluding remarks and outlook for possible future targets of speech

T. Ganchev, *Contemporary Methods for Speech Parameterization*, SpringerBriefs in Electrical and Computer Engineering, DOI 10.1007/978-1-4419-8447-0_1,
© Springer Science+Business Media, LLC 2011

parameterization research. Finally, Sect. 9 provides some links to other sources of information and to publically available software, which offer ready-to-use implementations of these speech features.

Keywords Speech pre-processing • Speech parameterization • Mel-scale • Critical bands • Cepstrum • Sub-band processing of speech • Time-frequency decomposition of speech • Cepstral analysis of speech • Speech features • Linear frequency cepstral coefficients • Mel frequency cepstral coefficients • Human factor cepstral coefficients • Perceptual linear prediction cepstral coefficients • Wavelet packet transform-based speech features • Wavelet packet features • Monophone recognition • Continuous speech recognition • Speaker recognition

1 Introduction

Section 1 offers background information about the origins of speech parameterization research and three important concepts, which lay the foundations of the contemporary short-time speech parameterization methods. Among them are (i) the sub-band processing of speech, (ii) the nonlinear perception of pitch and the critical band concept, and (iii) the short-time cepstrum of speech. We consider this background information an important addition to the subsequent technical sections as it contributes not only for the better understanding of the common grounds of the present-day frame-based speech parameterization techniques but also for clarifying their dissimilarities.

1.1 Speech Front-End

Speech parameterization is an important processing step in nearly all contemporary applications of speech technology, including the ubiquitous mobile phones, voice banking services, voice-driven information services, etc.

In brief, speech parameterization aims to calculate a small set of speech parameters, which well describe the properties of a given portion of the speech signal. Next, these speech parameters, which are also often referred to as speech features, act as input data to a certain machine learning technique, so that the linguistic and extralinguistic[1] information carried by the speech signal can be recognized and interpreted for the needs of human–machine interaction, or so that the speech can be reconstructed for the needs of speech communication among humans.

---

[1] Here extralinguistic information stands for any physiological or behavioral characteristics that can be inferred from the acoustic speech signal, such as: speaker identity, gender, age, height, body size, affective state, etc.

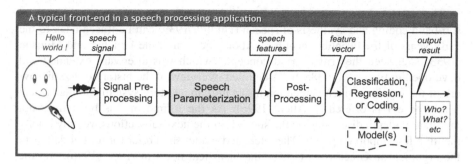

Fig. 1.1 Block diagram of a typical front-end, commonly used in speech technology applications

The general structure of a speech processing front-end, which is commonly used in speech technology applications, is illustrated in Fig. 1.1. As the figure shows, once captured, the speech signal is firstly pre-processed and after that it becomes subject to speech parameterization. The speech parameterization process converts the input speech into sequences of speech features which after some post-processing form a sequence of feature vectors. Each feature vector describes the properties of a small quasi-stationary portion of the speech signal, referred to as speech frame. A typical post-processing of the speech features aims at changing the statistical properties of the initially computed speech parameters and/or the size of the feature vector, and at gaining some advantage in terms of performance in the subsequent classification, regression, or encoding stage.[2] The proper choice of speech pre-processing, speech parameterization, and speech feature post-processing is crucial for the correctness of the outcome and for the performance of the entire system.

1.2 The Origins of Speech Parameterization

The contemporary speech parameterization techniques discussed in this book evolved from the merge of three main concepts, which were developed during the past century of speech processing research, and namely:

(i)   The idea for sub-band processing of speech
(ii) The understanding about the nonlinear pitch perception in the human auditory system and the critical band concept
(iii) The concept of cepstral analysis of speech

---

[2] The same signal pre-processing, speech parameterization, and post-processing steps are an indispensable part of the model(s) development process (not shown in Fig. 1.1).

The sub-band analysis of speech[3] (Miller 1916; Crandall 1917, 1925) and sub-band coding and synthesis of speech (Dudley 1939) can be traced back to the early years of the twentieth century.[4] About the same time Fletcher and Munson (1933) suggested the critical band concept,[5] which was afterward extended and developed in Fletcher (1938a, b). A detailed overview of the history of the discovery of the nonlinear pitch perception in the human auditory system and the critical band concept is offered in Allen (1996). As the format of this book does not facilitate in-depth discussion on the subject, in the next subsections we only briefly mention few important points. The interested reader shall refer for further details to the argument in Greenwood (1991), the historical notes in Allen (1996), and the related discussions in the online *Auditory list archives* (Research in Auditory Perception) hosted at the McGill University Web site.[6]

According to Noll (1967), the third important development, that is, the concept of cepstral analysis, was first suggested by Tukey toward the end of 1959. In brief, Tukey advised Bogert on how to deal with periodic ripples in the spectra of seismic waves, which he was investigating at that time, and following that advice Bogert started[7] experimenting with the cepstral analysis technique at the beginning of 1960. The results from this work were reported in Bogert et al. (1963), where the power cepstrum was introduced.

Again, according to Noll (1967), in June 1962, Schroeder, who was aware of the development of the power cepstrum, suggested to Noll the use of short-time cepstrum analysis for the needs of pitch estimation of speech. The results from this study were first published in full text (Noll 1964) in February 1964, and then reported at the 67th meeting of the Acoustical Society of America in May 1964 (Noll and Schroeder 1964). Thus, the article of Noll (1964) was the first publication, where the short-time cepstral analysis of speech was employed for the needs of speech parameterization. Later on, inspired by a subsequent work of Noll (1967), Oppenheim and Schafer (Oppenheim 1967; Oppenheim and Schafer 1968a) generalized the cepstral analysis technique and defined a more general case, where the cepstrum accounts for both the magnitude and phase spectrum.

---

[3] Although in the early years of the twentieth century Fourier analysis (sometimes involving measurements by hand!) was already accepted as a method for sound analysis, it was Miller (1916), who pioneered the systematic study of the energy distribution of speech. At about the same time Crandall (1917), also started systematic estimation of the energy distribution in speech, but more importantly, he started experimenting with manipulating certain frequency sub-bands and studied their effect on speech reproduction. Comprehensive results on the energy distributions of various speech sounds are available in Crandall (1925).

[4] A brief account of the main research trends on acoustic phonetics in the USA during the twentieth century is available in Mattingly (1999).

[5] In the speech recognition community the critical band concept is often credited to a subsequent overview article of Fletcher (1940).

[6] Auditory list archives, http://lists.mcgill.ca/archives/auditory.html

[7] From the distance of time, Bogert himself commented (Bogert 1967) that the work on the cepstrum started in 1962, but perhaps he meant the work on the paper Bogert et al. (1963).

In order to emphasize the difference with the power spectrum, which only accounts for the magnitude spectrum, Oppenheim and Schafer introduced the term *complex cepstrum*, as their generalized cepstrum uses the complex Fourier transform and the complex logarithm.

Although at present it is the power cepstrum that is generally employed in the speech feature extraction process, including all speech parameterization methods considered in this book, in the speech processing community, the series of publications of Oppenheim and Schafer (Oppenheim 1967; Oppenheim and Schafer 1968a, b; Oppenheim et al. 1968) are widely regarded[8] as the introduction of the cepstral analysis for the needs of speech parameterization. The interested reader shall refer to Deller et al. (1993) for a detailed discussion on the properties of the power cepstrum (Bogert et al. 1963; Noll 1964) and the more general complex cepstrum (Oppenheim and Schafer 1968a; Oppenheim et al. 1968).

1.3   The Technical Mel-Scale

The *Mel* (from *Mel*ody) scale is a heuristically derived perceptual scale that provides the relation between perceived frequency (referred to as pitch) of a pure tone as a function of its acoustic frequency. The definition of the Mel scale was firstly formulated by Stevens et al. (1937). In brief, Stevens et al. organized experiments in which subjects were required to adjust the frequency of a stimulus tone to be half as high as that of a comparison tone. Based on the experimental results Stevens et al. proposed the unit Mel for that scale equal to 1/1000 of the reference point, selected at 1000 Mels. The reference point between this scale and normal frequency measurement is defined by equating a 1000 Hz tone, 40 dB above the listener's threshold, with a pitch of 1000 Mels. Above the frequency of about 500 Hz, larger and larger intervals are judged by listeners to produce equal pitch increments. As a result, four octaves on the Hertz scale above 500 Hz are judged to comprise about two octaves on the Mel scale. Later on, in Stevens and Volkman (1940), the original Mel scale was revised. In this later work, the authors presented results from a second experiment with a modified experimental setup, which resolves some differences among the test subjects. The updated version of the Mel scale is illustrated in Fig. 1.2. Equal increments in Mels correspond to equal increments of the perceived pitch of pure-tone stimuli.

There exist various approximations of the nonlinear pitch perception of the human auditory system. An early approximation, referred to as the Koenig scale (Koenig 1949), is exactly linear below 1000 Hz and logarithmic above 1000 Hz.

---

[8] From the distance of time Oppenheim and Schafer (2004) seem to favor the presentation in Oppenheim et al. (1968) as a reference to the complex cepstrum. However, a presentation by Oppenheim (1967) and a manuscript submitted in September 1967 (Oppenheim and Schafer 1968a) seem to precede in time the submission of the work (Oppenheim et al. 1968).

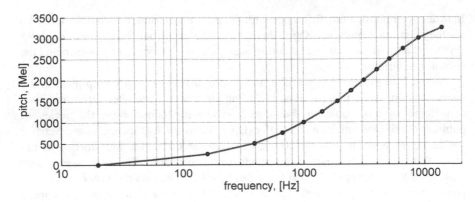

Fig. 1.2 The technical Mel scale. The translation from frequency to Mels is reconstructed according to the tabulation given by Beranek (1949)

It provides a computationally inexpensive representation of the Mel scale, which however is not very precise for frequencies both lower and higher than 1000 Hz. The values for the Koenig scale are calculated (Miller 1989) as:

$$
\begin{aligned}
K_{lin} &= 0.002 f_{lin} & \text{for} & \quad 0 \leq f_{lin} \leq 1000 \\
K_{log} &= 4.5 \log_{10}(f_{lin}) - 11.5 & \text{for} & \quad 1000 \leq f_{lin} \leq 10000
\end{aligned}
\tag{1.1}
$$

$$
f_K = 500 \left[\frac{K_{lin}}{K_{log}} \right],
\tag{1.2}
$$

where (1.2) was suggested (Miller 1989) as corrective transformation so that the Koenig scale has the same reference point at 1000 Hz as the other approximations. A more precise approximation of the Mel-scale was suggested by Fant (1949), in the general form:

$$
\hat{f}_{mel} = k_{const} \cdot \log_n \left(1 + \frac{f_{lin}}{F_b} \right),
\tag{1.3}
$$

for $F_b = 1000$. In the specific form presented in Fant (1973):

$$
\hat{f}_{mel} = \frac{1000}{\log_n 2} \cdot \log_n \left(1 + \frac{f_{lin}}{1000} \right)
\tag{1.4}
$$

this formula was found to represent a more close approximation of the available measurements (only for the frequency range of [0, 5000] Hz) when compared with the approximation offered by the Koenig scale. In addition, the formulation (1.4) is particularly interesting since the values of $\hat{f}_{mel}$ remain unaffected by the choice of the base n of the logarithm. Other versions of the Mel scale that were derived from (1.3) make use of the natural or of the decimal logarithm function, which results to a different choice of the constant k_{const}. The following two representations:

$$\hat{f}_{mel} = 2595 \cdot \log_{10}\left(1 + \frac{f_{lin}}{700}\right) \tag{1.5}$$

and

$$\hat{f}_{mel} = 1127 \cdot \log_e\left(1 + \frac{f_{lin}}{700}\right) \text{ aka } \hat{f}_{mel} = 1127 \cdot \ln\left(1 + \frac{f_{lin}}{700}\right) \tag{1.6}$$

are widely used in the various implementations of the MFCC. Compared to (1.4), the formulae (1.5) and (1.6) provide a closer approximation of the empirical data for frequencies below 1000 Hz, at the price of higher inaccuracy for frequencies higher than 1000 Hz. A comprehensive comparison of various approximations of the nonlinear pitch perception is available in Umesh et al. (1999). For a further discussion on the Mel scale, the interested reader could benefit from the critical account offered in Greenwood (1997).

1.4 The Critical Band Concept

Although other earlier studies were present in the first half of the twentieth century, a considerable progress in the exploration of the human auditory system was made by Fletcher (1938a, b, 1940). Based on experiments (Fletcher and Munson 1933), which measure the threshold of hearing for a sinusoidal signal as a function of the bandwidth of a band-pass noise masker, Fletcher (1938a, b) suggested that the peripheral auditory system behaves as if it consisted of a bank of band-pass filters with overlapping pass-bands. Nowadays, these filters are referred to as *auditory filters*. Fletcher relied on the assumption that the frequency selectivity of the auditory system and the characteristics of its corresponding auditory filters can be investigated by conducting perceptual experiments based on the technique of masking. As it is now widely known, the masking effect decreases the sensitivity of the human auditory system to the detailed spectral structure of a sound within the bandwidth of a single auditory filter. To describe the effective bandwidth of the auditory filter over which the main masking effect takes place, Fletcher introduced the term of *critical bandwidth*, and then used the phrase *critical bands* to refer to the concept of the auditory filters.

Since Fletcher's first description of the critical bands, many experimenters attempted to estimate it. Zwicker et al. (1957), and later on Zwicker (1961), estimated that the critical bandwidth is constant and equal to 100 Hz for frequencies below 500 Hz, while for higher frequencies it increases approximately in proportion with the center frequency. It has to be mentioned here that the critical bandwidth relationship derived by Zwicker, was estimated when there were only few estimates available for low center frequencies and thus Zwicker had to use extrapolation. More important, however, is that the work of Zwicker demonstrated how an objectivist theory of pitch should be developed in contrast to the subjectivist view that led to the Mel scale. In Zwicker and Terhardt (1980), the authors derived a function from frequency to subjective pitch, expressed in *Barks*, unit named after Heinrich *Bark*hausen, a German physicist. Despite the advantage of the Bark scale, namely, of being laid on solid theoretical grounds, nowadays both the Mel and Bark scales remain in use.

In more recent experiments, Glasberg and Moore (1990) demonstrated that the estimated by Zwicker critical bandwidths are not accurate, and the critical bandwidth continues to decrease with frequency yet for frequencies far below 500 Hz. In Moore (2003), an analysis of various attempts for determining the shape of the auditory filters and estimating the Equivalent Rectangular Bandwidth (ERB) demonstrated that there are discrepancies between the present understanding of the critical bandwidth and the models developed by Zwicker.

The ERB might be regarded as a measure of the critical bandwidth, and according to Moore (2003), it is equal to the bandwidth of a perfect rectangular filter, whose pass-band is equal to the maximum transmission of the specified filter and transmits the same power of white noise as the specified filter. Equation 1.7 presents the ERB as a function of center frequency f using moderate sound levels for young people with normal hearing (Glasberg and Moore 1990):

$$\text{ERB} = 24.7 \left(4.37 \frac{f}{10^3} + 1 \right), \tag{1.7}$$

where the values of ERB and f are specified in Hz. As presented in Moore (2003), Eq. 1.7 fits roughly the values of ERB estimated in many different laboratories. Therefore, ERB approximates the critical bandwidth, which in turn is a subjective measure of the bandwidth of the auditory filters.

1.5   A Brief Note on Short-Time Speech Parameterization

The great diversity of speech parameterization techniques observed today is due to the continuous effort for improving the speech feature extraction process through the incorporation of the latest insights from the area of psychoacoustics. In the following, we mention some more interesting speech parameterization techniques that attracted attention during the years and that were used on various speech processing tasks.[9]

---

[9] Here, we did not intend an exhaustive enumeration of the great number of speech parameterizations that were reported beneficial over the commonly used speech features as this is beyond the scope of this book.

In the early times after the introduction of the short-time cepstral analysis of speech, various speech parameterizations dominated the speech recognition area: Real Cepstral Coefficients (RCC) introduced by Oppenheim and Schafer (1968a), Linear Prediction Coefficients (LPC) proposed by Atal and Hanauer (1971), Linear Predictive Cepstral Coefficients (LPCC) derived by Atal (1974), and the Mel Frequency Cepstral Coefficients (MFCC) formulated in Bridle and Brown (1974) and Davis and Mermelstein (1980).

In Davis and Mermelstein (1980), and later on in other studies, it was demonstrated that the MFCC outperform the LPC, LPCC, and other speech features on the task of speech recognition. From a perceptual point of view, the MFCC roughly resemble the properties of the human auditory system, since they account for the nonlinear nature of pitch perception, as well as for the nonlinear relation between intensity and loudness. These biologically plausible characteristics make the MFCC more suitable to the needs of speech recognition than other formerly used speech parameters like RCC, LPC, and LPCC. This success of MFCC, combined with their robust and cost-effective computation, turned them into the typical choice of speech parameterization in nearly all speech recognition applications. Due to their success and the widespread use in the speech recognition community, the MFCC became also widely used on speaker recognition tasks, although they might not represent the individuality of human voices with sufficient accuracy. In fact, when MFCC are used for speech recognition, it is feasible to suppress the distinctiveness of the different voices,[10] while the linguistic information remains unaffected by this process.[11]

In the past two decades, the consequent advances in the understanding of the human auditory system reflected in the emergence of new speech parameterization techniques. Many innovative speech features were designed, such as: the Perceptual Linear Prediction (PLP) cepstral coefficients proposed in Hermansky (1990), the Adaptive Component Weighting (ACW) of Assaleh and Mammone (1994a, b), etc. but the MFCC paradigm preserved its predominance. As a matter of fact the PLP cepstral coefficients also made their way in the speech recognition community and are occasionally used instead of the MFCC.

In addition, many of the relatively more recent speech parameterization schemes, such as the various wavelet-based features (Sarikaya et al. 1998; Sarikaya and Hansen 2000; Farooq and Datta 2001; Siafarikas et al. 2005, 2007; etc.), although presenting reasonable alternatives for the same tasks, did not gain a widespread practical use. The last is primarily due to the poor understanding of the advantages they offer on certain speech processing tasks, their relatively more sophisticated computation,

---

[10] For instance, through the application of a certain vocal tract length normalization technique.

[11] However, in the text-independent speaker recognition task, the presence of linguistic information in the short-time speech features is not beneficial. In fact, its presence makes the speaker recognition process even more difficult, as the linguistic information constitutes additional source of variability, which is not related to the main objectives of the text-independent speaker recognition tasks.

and to the lack of open-source implementations that could promote their use in the speech processing community.

To this end, wavelets have been employed in cepstrum-based schemes for speech parameterization in two different ways. The first approach makes use of wavelet transform as an effective decorrelator of the speech features instead of the discrete cosine transform (Tufekci and Gowdy 2000; Sarikaya and Hansen 2000). According to the second approach, wavelet transform is applied directly on the speech signal for the needs of time-frequency analysis of the signal. In this second case, either wavelet coefficients with high energy are taken as speech features (Long and Datta 1996), which nonetheless suffer from shift variance, or sub-band energies are computed instead of the Mel filter-bank sub-band energies as in Sarikaya et al. (1998), Sarikaya and Hansen (2000), Farooq and Datta (2001), Siafarikas et al. (2005, 2007), etc.

In particular, in the speech recognition area, the wavelet packet transform, employed for the computation of the spectrum, was first proposed in Erzin et al. (1995). Later on, wavelet packet bases were used in Sarikaya et al. (1998), Sarikaya and Hansen (2000), and Farooq and Datta (2001, 2002) for the construction of speech features that are based on close approximations of the Mel-frequency division using Daubechies' orthogonal filters with 32 and 12 coefficients, respectively. In addition, in an attempt to derive speech features that are optimized for the speaker recognition task Siafarikas et al. (2004, 2005, 2007), performed an objective evaluation of various wavelet packet trees for speech decomposition, and systematically selected the most beneficial ones. These studies resulted in speech parameterizations that use filter-banks with double or triple the number of sub-bands when compared to the traditional MFCC implementations.

Recently, Nogueira et al. (2006) studied three speech decompositions that are based on the Advanced Combinational Encoder (ACE) "NtoM" strategy (Nogueira et al. 2005). Specifically, Nogueira et al. (2006) investigated the appropriateness of three different basis functions, namely, the Haar wavelet, the Daubechies' wavelet of order 3 and the Symlets family of wavelets, for improving the speech intelligibility in cochlear implants. All experiments were performed on a common decomposition tree that closely follows the frequency bands associated with the electrodes in the ACE strategy.

Besides the wavelet-based speech features in the past decade a promising new data-driven paradigm for speech parameterization, known as TRAP-TANDEM,[12] emerged (Hermansky 2003). However, although the TRAP-TANDEM paradigm attracted some attention in the speech recognition community (Athineos et al. 2004; Grezl et al. 2004; Valente 2010), at present the traditional short-time frame-based parameterization of speech dominates in virtually all kinds of speech processing tasks.

---

[12] The interested reader is encouraged to refer to Hermansky (2003), not only for a comprehensive description of the TRAP technique, but also for a succinct analysis of the shortcomings of the contemporary speech parameterizations.

Finally, we cannot neglect the fact that the common availability of significant computational resources fostered the use of speech feature vectors, which consist of hundreds or thousands of speech parameters. For instance, nowadays, in the speaker recognition community, the use of supervectors of features is a common practice.[13] Due to the recent emergence of the open-source openSMILE audio parameterization (Eyben et al. 2010), the use of hundreds of speech/audio parameters also becomes a reality in the research on affect, emotion, age, gender recognition (Schuller et al. 2009, 2010; Batliner et al. 2011), speaker height estimation from speech (Mporas and Ganchev 2009; Ganchev et al. 2010) etc. However, as most of these recent trends directly build on the use of the traditional short-time speech parameterizations, we focus our attention solely to a number of the aforementioned short-time DFT- and DWPT-based speech parameterization techniques that operate on the level of individual speech frames. The understanding of the benefits that these short-time speech parameterizations offer and their inherent limitations may contribute to further advances in the use of large speech feature supervectors, or to the meticulous selection of subsets of speech features, which offer certain performance trade-offs.

2 Speech Pre-processing

Section 2 offers details on the purpose of speech pre-processing and discusses two pre-processing steps, namely, the speech pre-emphasis and windowing, which are commonly used in the short-time frame-based speech parameterization methods. In brief, Sect. 2.1 recapitulates the basics of speech pre-processing. Furthermore, Sect. 2.2 summarizes the sequence of speech pre-processing steps that are used in the DFT- and the DWPT-based speech parameterization techniques discussed in Sect. 3 and 4, respectively.

2.1 Basics of the Speech Pre-processing

The speech pre-processing stage is an essential part of any speech technology application as it prepares the speech signal in a format that facilitates the speech parameterization process. In laboratory experiments with pre-recorded databases of digitized clean speech, the pre-processing could be as simple as:

(i) Performing pre-emphasis of the speech signal (i.e., decreasing the relative contribution of the low-frequency components so that the upper frequency components become more prominent)

---

[13] Details are available in a recent overview of the speaker recognition technology by Kinnunen and Li (2010).

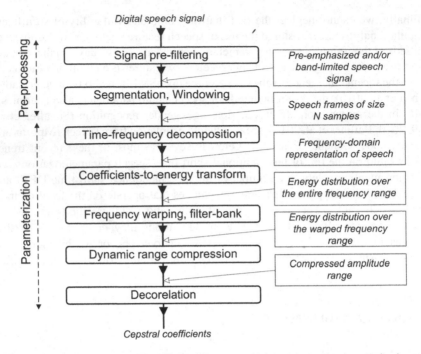

Fig. 2.1 A general block diagram of the speech pre-processing and speech parameterization steps for the short-time speech features considered in this book

(ii) Subsequent framing to form the sequence of short-time segments that will be parameterized next

However, in real-life applications, the speech pre-processing also includes all efforts for sound capturing, signal amplification, band-limiting, (re)sampling, speech enhancement, etc. As these pre-processing steps mostly depend on the operational conditions and on the specifics of each real-life application, in the following we mainly focus on the mandatory speech pre-processing steps that are considered an integral part of nearly all short-time frame-based speech parameterization methods.

A general block diagram that summarizes the processing steps for most[14] of the speech parameterization techniques considered in this book is shown in Fig. 2.1. As Fig. 2.1 shows, we assume that the speech signal is already in digital format and acts at the input of the speech pre-processing stage. The first pre-processing step, denoted as *signal pre-filtering*, is often implemented as some form of band-pass filtering of the speech signal. This could be seen as a simple but often sufficient noise reduction pre-processing, which aims at improving the signal-to-noise ratio (SNR)

---

[14] The only exception among the speech parameterizations considered in this book is the computation of the perceptual linear prediction (PLP) cepstral coefficients (Hermansky 1990), which follows a different sequence of processing steps.

of speech before parameterization. In the simplest case of additive low-frequency noise (for instance, from the wind, from a car engine, a cooling fan, or from vibrations in a power-supply unit), such a band-limitation attenuates the undesirable low-frequency components in the signal, and thus contributes for the improvement of the overall SNR.

More importantly however, the band-pass filtering is quite important in telephone speech applications where overdriven speech, that is, nonlinearly distorted speech signal with clipping by amplitude, is a common phenomenon.[15] In order to reduce the spectral distortions from aliasing in the spectrum, which the overdriving of speech causes, it is advisable[16] that the amplitude of the already overdriven speech signal is first attenuated by 3 dB and afterward the overdriven signal is passed through a band-pass filter. Except for the attenuation of the dc-offset and low-frequency interferences, the band-pass filtering also reduces the contribution of the high-frequency components which are due to the sharp edges of the clipped signal. The last prevents the introduction of phantom low-frequency components in the estimation of the spectrum, that is, aliasing, which otherwise the clipping of signal causes.

As a part of the signal pre-filtering step, we also consider the high-pass filtering of speech, referred to as speech pre-emphasis. The pre-emphasis of speech is typically performed before the speech parameterization process[17] as it decreases the relative contribution of the low-frequency components in voiced speech, and thus makes the higher frequency components more prominent. Pre-emphasis of speech is needed as the low-frequency formants of speech have much higher energy when compared to the upper-frequency formants (especially the fourth and fifth formants), and this renders difficult the accurate estimation of the position and the bandwidth of the upper formants. In order to cope with this difficulty, the pre-emphasis of speech is used to compensate the slope of -6 dB per octave in the spectrum of voiced speech,[18] and thus decreases the dynamic range of the amplitudes of the speech formants. The last facilitates the proper estimation of the upper formants and the more adequate modeling of the upper part of the speech spectrum. For the unvoiced portions of speech the pre-emphasis filtering is not necessary but it is often performed for the sake of simplification of the signal pre-processing stage.

---

[15] The overdriven telephone speech most often occurs due to the close position of the microphone to the mouth of the speaker, the differences in the temperament and the speaking style among speakers, the variations in the affective state of the speaker, changes in the speaking style in the presence of acoustic interferences (Lombard effect), etc.

[16] This author advocates band-pass filtering as a simple but effective and rewarding way to deal with overdriven telephone speech. A fifth-order Butterworth band-pass filter with cut-off frequencies $0.01 \div 0.03 \times f_s$ and $0.475 \times f_s$, where f_s is the sampling frequency, usually helps.

[17] An exception here is the computation of the PLP cepstral coefficients, where the pre-emphasis is performed in the frequency domain.

[18] The slope of -6 dB per octave in the spectrum of voiced speech is due to the combined effect of the slope -12 dB per octave due to the excitation source and the slope of $+6$ dB per octave due to the sound radiation through the lips – details are available in (Fant 1956).

The pre-emphasis filter is often defined with its transfer function,

$$H(z) = 1 - az^{-1}, \tag{2.1}$$

where the single filter coefficient a, which usually takes value in the range $0.95 \div 0.97$, controls the cut-off frequency of the filter and thus the degree of attenuation of the low-frequency components in the voiced speech. When the pre-emphasis filter is applied both on voiced and unvoiced speech parts, the filter-coefficient might receive a slightly lower value as a trade-off between the distortion of the unvoiced speech and the needed pre-emphasis of the formants. The pre-emphasis filter also reduces any drift and low-frequency deformations of the speech signal, which may arise from air-flow bursts directed toward the microphone (deep breathe, laughter, etc.).

Finally, the signal pre-filtering step often includes a certain more advanced noise reduction technique, which has to deal with the interferences from the operational environment.

In order to better understand the second speech pre-processing step, often referred to as *speech framing* or *windowing*, let us first have a look at the subsequent step 3 in Fig. 2.1, which is the time-frequency decomposition of speech. In brief, the time-frequency decomposition of speech,[19] defined at step 3, aims at obtaining the frequency content of the signal for subsequent instants in time, so that sound events can be localized both in time of appearance and in frequency composition. When using the discrete Fourier transform, this is achievable by giving emphasis to a certain time interval and de-emphasizing the rest of the signal before performing the frequency analysis. Like many other quantity estimation methods, the time-frequency decomposition also obeys Haisenberg's uncertainty principle for observable physical quantities, so that the achievable time resolution and frequency resolution are traded off one for another.

For instance, in the case of the short-time discrete Fourier transform (stDFT), for a fixed sampling frequency, the length of the time interval of interest, often referred to as *window*, bounds both the frequency resolution and the time resolution of the decomposition. This window has to be long enough to guarantee sufficient frequency resolution for detecting the required level of details in the frequency domain but not too long as speech is a nonstationary process and the signal within this interval has to preserve its statistical properties unchanged. Thus, the window has to be sufficiently short for discriminating among different events in the time domain.

In order to clarify the role of the abovementioned window, let us denote with n the discrete-time index, and with $x(n)$ a discrete-time speech signal[20] that has been

---

[19] This concept was initially developed in D. Gabor (1946) Theory of Communication. *Journal of Institution of Electrical Engineers* 93(3):429–457, November 1946.

[20] In this book we discuss only discrete-time signals and therefore we can take the liberty to use round brackets instead of the strict squire brackets, which are commonly used in the designation of discrete-time series. The author hopes that this simplification will not introduce inconvenience and will not put the reader to confusion.

sampled with sampling frequency f_s. Applying the stDFT on the signal $x(n)$ is defined[21] as:

$$X(l,k) = \sum_{i=0}^{N-1} x(i+l) \cdot W(i) \cdot \exp\left(-\frac{j2\pi ik}{N}\right), \quad 0 \le i, k \le N-1. \quad (2.2)$$

Here, i is the index of the time domain samples, k is the index of the Fourier coefficients $X(l,k)$, and l denotes the relative displacement from the beginning of the signal. The window $W(.)$, which has length of N nonzero samples, provides the local properties of the stDFT and is considered to have value zero outside this time interval, that is, $W(i)>0$ for $0 \le i \le N-1$ and $W(i) = 0$ otherwise. Thus, the window $W(.)$ selects from the entire signal only a portion of N samples, which is subject to frequency analysis. Depending on the value of l, with $l = mT, m = 0, 1, ...$, where T is the number of samples to skip from a frame to the next one, two consecutive windows may overlap or may not. The successive recomputing of (2.2) for increasing values of l is equivalent to transforming the discrete-time speech signal $x(n)$ into a sequence of short quasi-stationary intervals of length N that overlap each other with $\max\{0, (N-T)\}$ samples.

With the above discussion, we clarified the purpose of windowing (framing) of speech, and at this point we refer back to the second speech pre-processing step in Fig. 2.1. As (2.2) shows, the window $W(.)$ and the speech signal $x(.)$ are multiplied in the time domain before the frequency analysis, which corresponds to convolution in the frequency domain, and therefore, the shape of the window heavily affects the proper estimation of the spectrum of the original signal.

In order to limit the negative effects caused by setting to zero of the signal outside the interval $0 \le i \le N-1$, the windowing function should be smooth and have values close to zero near the border points in the time domain, and should have the properties of the perfect low-pass filter in the frequency domain. The interested reader may wish to follow the exposition in Harris (1978) and Nuttall (1981) for a comprehensive account on the use of various windowing functions for harmonic analysis with the stDFT. In this book, we only consider two types of windows: (i) rectangular window (aka Dirichlet window) and (ii) the Hamming window, as they are used in the computation of the speech parameterizations discussed in Sects. 4 and 3, respectively.

As shown in Fig. 2.2 with dashed line, the rectangular window has uniform unit weighting for the entire interval of interest and value zero outside, that is,

$$W(i) = \begin{cases} 1 \text{ for } 0 \le i \le N-1 \\ 0 \text{ for any other } i \end{cases}, \quad (2.3)$$

---

[21] Here and in Sect. 3 of this book, the shift of the summation boundaries with $N/2$ samples and the resulting shift of the phase spectrum are not considered as a reason for a major worry, as the phase information is not accounted in the computation of the speech parameterizations based of this specific version of the stDFT.

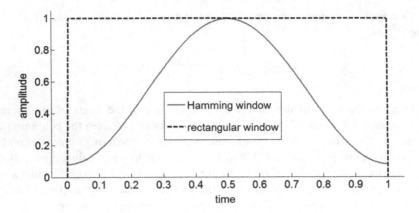

Fig. 2.2 The Hamming window (*solid line*) and the rectangular window (*dashed line*)

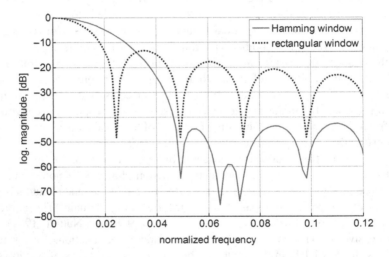

Fig. 2.3 The logarithmically compressed power spectrum for the Hamming window (*solid line*) and the rectangular window (*dotted line*) for window length $N = 256$ samples

where N is the desired length of the window. For a fixed value of N, the rectangular window has the largest effective time-domain width among all other window functions. The rectangular window also allows preserving the energy of the signal unchanged. However, due to the abrupt change from value one to value zero at the borders of the window, in the frequency domain its frequency transform does not comply well with the desired perfect low-pass filter – the first sidelobe is only 13 dB below the main lobe (Fig. 2.3). By that reason, the rectangular window introduces leakage of energy through the sidelobes of its Fourier transform. In this book, the rectangular window is only used in the speech pre-processing for the speech parameterizations based on the discrete wavelet packet decomposition, as they effectively weight the signal on their own through the wavelet functions.

As shown in Fig. 2.2, the Hamming window has an elevated symmetric bell-like shape and offers a smooth attenuation of the signal toward the boundary points.

The slight elevation of the bell end-points above the value zero offers a trade-off between the effective width of the Hamming window in the time domain and the amplitude of the sidelobes of its frequency domain transform. The Hamming window is defined as:

$$
W(i) = \begin{cases} 0.54 - 0.46\cos\left(\dfrac{2\pi i}{N}\right), & \text{for } i = 0, 1, ..., N-1, \\ 0 & \text{for any other } i \end{cases}
\tag{2.4}
$$

where N is the window size. When compared to the rectangular window, the Hamming window reduces the spectral distortions in the frequency domain as it softens the abrupt change to value zero at the boundary points. As Fig. 2.3 shows, the Hamming window has a smaller leakage of energy through the sidelobes as the first sidelobe is at approximately 44 dB below the main lobe, while for the rectangular window, the first sidelobe is at −13 dB. However, the smaller leakage of energy for the Hamming window is traded off for a twice wider main lobe, which is equivalent to twice worse frequency resolution when compared to the rectangular window.

Concentrating most of the energy in a narrow main lobe is a highly desirable property as the estimation of the amplitude of the frequency components in the spectrum is a sum of all terms in (2.2). Therefore, any nonperfect low-pass behavior of the window $W(.)$, such as a nonuniform pass-band transmission, a significant energy leakage through the sidelobes, a wide main lobe, would result in a less accurate representation of the true spectral content of the original signal, which consequently results to speech features that are less accurate representation for the specific speech frame.

With this brief note on the two mandatory speech pre-processing steps – (i) pre-filtering (step 1) and (ii) the windowing (step 2) – we conclude our discussion on the speech pre-processing stage (refer to Fig. 2.1). As the signal processing after step 2 is dissimilar for the speech parameterization techniques discussed in Sects. 3 and 4, we will not discuss them in detail in this short introduction. Instead, in the following section, we summarize the speech pre-processing steps used in the speech parameterizations methods discussed in Sects. 3 and 4.

2.2 Summary of the Speech Pre-processing Steps

The pre-processing of the time domain speech signal for the needs of DFT- and DWPT-base speech parameterization methods (Sects. 3 and 4) consists of the following steps:

(i) Mean value removal, which aims at eliminating the dc-offset that might have occurred during signal acquisition, and adjusting the signal amplitude to the desired dynamic range [−1, 1].

(ii) Pre-emphasis of the signal for compensating the -6 dB slope in the spectrum of voiced speech.
(iii) Windowing with a Hamming window (for the needs of the speech parameterization methods in Sect. 3) or with a rectangular window (for the needs of the speech parameterization methods in Sect. 4).

A succinct description of these steps is provided in the following. As before, let us denote with n the discrete-time index, and with $x(n)$a discrete-time speech signal that has been sampled with sampling frequency f_s. The dc-offset (the mean value) removal and the adjustment of the signal amplitude can be written as:

$$\widehat{x}(n) = g_{AGC}(x(n) - \mu_x),\qquad(2.5)$$

where μ_x is the mean value of $x(n)$, and g_{AGC} is an adjustable gain factor which keeps the input signal in the range $x(n) \in [-1, 1]$.

The first-order pre-emphasis filter, defined through its difference equation as

$$y(n) = \widehat{x}(n) + a\widehat{x}(n - 1)\qquad(2.6)$$

is applied on the zero-mean signal $\widehat{x}(n)$, so that the low frequencies are suppressed. In order (2.6) to behave as a high-pass filter, its only adjustable coefficient, a, has to receive a negative value. The actual value of a, which depends on the sampling frequency, f_s, and on the desired cut-off frequency, f_{cut}, of the filter is computed as

$$a = \exp\left(-\frac{2\pi f_{cut}}{f_s}\right)\qquad(2.7)$$

and is usually in the range $a \in [-0.97, -0.95]$.

Next, the speech framing is performed with the rectangular window (2.3) or with the Hamming window (2.4), for all feasible values of $m = 0, 1, ..., L - 1$. This is equivalent to the conversion of the pre-emphasized signal $y(n)$ to a $L \times N$ matrix, where each row is one frame of the speech signal weighted by the window function

$$s(i, l) = y(i + l)W(i), \quad i = 0, 1, ..., N - 1, \quad l = mT, \quad m = 0, 1, ..., L - 1,\quad(2.8)$$

and the number of rows L depends on the length of the speech signal, the skip time T, and to a smaller degree to the window's width N.

As the speech parameterization techniques that we consider in the following sections process each speech frame independently from its neighbors, for simplicity of exposition, we drop the index l. Therefore, in the following sections, we will often speak about the discrete signal $s(n)$ defined in the interval $n = 0, 1, ..., N - 1$, where $s(n) \equiv s(i, l)$ for a given fixed value of m and $i = 0, 1, ..., N - 1$. Wherever needed we will clarify that the signal $s(n)$ has been subject of the speech pre-processing steps (2.5) and (2.6) described in this section, and has been weighted (2.8) with the rectangular window (2.3) or with the Hamming window (2.4) function.

3 DFT-Based Speech Parameterization

The twentieth century was marked with significant advances in the field of psycho-acoustics and great improvements in our understanding of the functioning of the human auditory system (Fletcher 1940; Zwicker 1961; Patterson and Moore 1986; Glasberg and Moore 1990; Moore and Glasberg 1996; etc.). Following this progress, the speech parameterization process was enriched with a new category of methods, which is commonly labeled as biologically inspired speech parameterization. This group of methods includes a diversity of speech parameterization techniques, which bear resemblance to various aspects of the human auditory perception. Among them are the various implementations of the Mel frequency cepstral coefficients (MFCC), Perceptual linear prediction (PLP) cepstral coefficients, Human factor cepstral coefficients (HFCC), etc. All these build on the three main speech processing concepts discussed in Sect. 1.2, namely, (i) the sub-band processing of speech, (ii) the nonlinear pitch perception in the human auditory system and the critical band concept, and (iii) the concept of cepstral analysis of speech.

In this section, we focus our attention to the various DFT-based speech parameter-izations, among which are three well-known implementations of the MFCC that are widely used in the contemporary speech technology applications, the linear frequency cepstral coefficients (LFCC), the PLP cepstral coefficients, and the recently proposed HFCC speech features. In the following subsections, we firstly offer a comprehensive description of each of these speech parameterizations and then discuss the main differences among them.

The idea for the MFCC, that is, applying the DCT[22] on the logarithmically compressed output of a filter-bank of filters with nonlinearly spaced center frequencies, and their use for the needs of speech recognition, is credited to Bridle and Brown (1974). However, it was the work of Davis and Mermelstein (1980), which further developed this concept and made the MFCC popular in the speech recognition community. Specifically, the applicability of MFCC to discriminate among phonetically similar words was studied in Mermelstein (1976). Later on, in Davis and Mermelstein (1980), the MFCC were reported advantageous over earlier DFT-based and linear prediction-based speech features on the task of monosyllabic word recognition. After that work of Davis and Mermelstein was published, the MFCC became widely used, and in the subsequent years, numerous variations and improvements of the initial idea were proposed. These variations differ mainly in the number of filters, the shape of the filters, the way the filters are spaced, the bandwidth of the filters, and the manner in which the frequency is warped. In addition to the aforementioned variables, the frequency range of interest, the selection of actual subset, and the number of MFCC coefficients that are employed in the subsequent recognition process can also be different. The application setup and the

---

[22] In Sect. 3.2 we clarify that principal component analysis was previously used in Klein et al. (1970) on the task of vowel identification and in Pols (1971) on the task of small vocabulary isolated word recognition, and it was already known (King 1971) that the principal component analysis and cosine transform are essentially equivalent with respect to their decorrelation effect.

objectives of the particular speech processing task also affect the choice of frame size, frame overlap, the coefficient of the pre-emphasis filter, windowing function, etc.

In all speech parameterizations discussed in Sect. 3 we make use of the common pre-processing[23] described in Sect. 2.2. In particular, we first overview the computation of the LFCC, four popular MFCC implementations, among which are the:

- MFCC-FB20 – as in Davis and Mermelstein (1980),
- HTK MFCC-FB24 – as in the Cambridge HMM Toolkit (HTK) (Young et al. 1995),
- MFCC-FB40 – as in the MATLAB Auditory Toolbox (Slaney 1998),
- HFCC-E-FB29 – as in Skowronski and Harris (2004),

widely credited as beneficial in the various speech processing tasks, and finally the PLP cepstral coefficients. These speech parameterizations differ mainly in the particular approximation of the nonlinear pitch perception of human, the filter-bank design, and the nonlinear function employed for compressing the dynamic range of the amplitude at the filter-bank output.

In the performance evaluations presented in Sects. 5–7, where various speech features are ranked on the tasks of monophone recognition, continuous speech recognition, and text-independent speaker verification, the widely-used HTK MFCC-FB24 and the MFCC-FB40 will be considered as intuitive reference baseline, against which the performance of the other speech parameterization are compared.

3.1   The LFCC

The linear frequency cepstral coefficients (LFCC) stem from the merge of the idea for sub-band processing of speech and the concept of cepstral analysis (a brief historical note is offered in Sect. 1.2). In fact, LFCC is a common designation for not just one but several implementations of the cepstral analysis of speech. From the 1960s to the 1980s, the cepstral coefficients were derived directly based on the spectrum of speech without applying a filter-bank before the DCT, and, for instance, this is the LFCC implementation in the widely cited work of Davis and Mermelstein (1980). Later on, the LFCC were given a new meaning, and at present, their computation involves applying a filter-bank of filters with linearly spaced center frequencies on the power spectrum before the logarithmic compression of the amplitude, as this seems to decrease their variability from the speaker-specific voice traits. In this section, we focus on these two ways for computing the LFCC parameters, which can be summarized as follows.

Let us denote with n the discrete time index, and with $x(n)$ a discrete-time speech signal that has been sampled with sampling frequency f_s, and therefore has spectral content bounded in the frequency range $[0, 0.5] f_s$. Let us consider that

---

[23] Except for the PLP analysis which applies the pre-emphasis filter in the frequency domain.

the signal $x(n)$ has been pre-processed as explained in Sect. 2.2, which resulted in the pre-emphasized, and weighted with the Hamming window signal $s(n)$, $n = 0, 1, ..., N - 1$ that corresponds to one speech frame of N samples. Next, each speech frame, $s(n)$, obtained in this manner becomes subject to the discrete Fourier transform (DFT),

$$S(k) = \sum_{n=0}^{N-1} s(n) \cdot \exp\left(\frac{-j2\pi nk}{N}\right), \quad k = 0, 1, ..., N - 1. \tag{3.1}$$

Here, n is the index of the time-domain samples, and k is the index of the Fourier coefficients $S(k)$. In the early years after the introduction of the cepstral analysis of speech, the cepstral coefficients were computed directly from the power spectrum, $|S(k)|^2$, of the signal $s(n)$, as (Davis and Mermelstein 1980):

$$LFCC(r) = \sum_{k=0}^{K-1} \log(|S(k)|^2) \cos\left(\frac{\pi rk}{K}\right), \quad r = 0, 1, ..., R - 1. \tag{3.2}$$

However, nowadays the LFCC defined with (3.2) are not used, and instead LFCC are estimated after applying a linear filter-bank on the power spectrum just before computing (3.2). This was observed to suppress the fine details in the cepstrum, and thus to reduce the sensitivity of LFCC to inter-speaker variability, a valuable advantage in the speaker-independent speech recognition applications.

In the early years of the twentieth century, researchers (Miller 1916; Crandall 1925; Dudley 1939) made use of ten linear sub-bands that covered the frequency range [0, 3000] Hz. However, with the advance of technology, the covered frequency range was extended and the number of filters increased. The last contributed to achieving a higher precision in speech analysis and speech reproduction. In accordance with research published in the early years of speech technology, and based on the critical band concept,[24] one can select a meaningful filter bandwidth in the range [100, 300] Hz, which offers an appropriate trade-off between frequency resolution and number of filters.[25]

---

[24] The critical band concept, as it was understood at that time, suggested a critical bandwidth of 100 Hz or lower for frequencies below 1000 Hz, and increasing by logarithmic law bandwidth for frequencies above 1000 Hz.

[25] In the two extreme cases, the number of sub-bands could be equal to the number of DFT coefficients in the frequency range $[0, 0.5] f_s$ or the entire bandwidth could be considered as one filter. As already mentioned, the former case corresponds to the cepstral coefficients defined with (3.2), which are sensitive to inter-speaker variability. In the latter case, we have just one sub-band, that is, no frequency content can be identified, and therefore we can compute only one cepstral coefficient, which is proportional to the total energy of the signal in that frame.

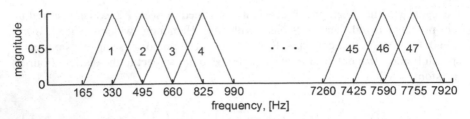

Fig. 3.1 A filter-bank of 47 triangular, equal-bandwidth and equal-height filters for the frequency range [165, 8000] Hz

In the following, we consider a linear filter-bank with equal-spaced filters each with bandwidth of 165 Hz. This bandwidth offers a decent balance between the frequency resolutions used at low and at high frequencies, and leads to the same number of filters as in the MFCC implementation[26] of Slaney (1998).

However, in the general discussion offered here, we adhere to the frequency range of interest [0, 8000] Hz, so that we keep uniform presentation with the other speech parameterizations considered in this book. Therefore, in order to cover this range with equal-width equally spaced sub-bands of 165 Hz, the filter-bank $H_i(k)$, $i = 1, 2, ..., M$, shall consists of $M = 48$ filters. Each of these filters is defined as:

$$H_i(k) = \begin{cases} 0 & \text{for} \quad k < f_{b_{i-1}} \\[2mm] \dfrac{(k - f_{b_{i-1}})}{(f_{b_i} - f_{b_{i-1}})} & \text{for} \quad f_{b_{i-1}} \le k \le f_{b_i} \\[2mm] \dfrac{(f_{b_{i+1}} - k)}{(f_{b_{i+1}} - f_{b_i})} & \text{for} \quad f_{b_i} \le k \le f_{b_{i+1}} \\[2mm] 0 & \text{for} \quad k > f_{b_{i+1}} \end{cases} \tag{3.3}$$

Here, the index i stands for the ith filter, f_{b_i} are the boundary points of the filters, and $k = 1, 2, ..., N$ corresponds to the kth coefficient of the N-point DFT. The boundary points f_{b_i} are expressed in terms of position, which depends on the sampling frequency f_s and the number of points N in the DFT. The centers of the linearly spaced filters are displaced 165 Hz one from another, and serve as boundary points for the corresponding neighboring filters. As the lowest frequency sub-band, [0, 165] Hz, contributes little to the speech content in many cases, it can be discarded without significant loss of information, and then we end up with $M = 47$ sub-bands. Figure 3.1 shows the equal-width equal-height filter-bank with 47 filters.

In the filter-bank version of the LFCC speech features, the DFT (3.1) is first applied and then the log-energy of the filter-bank outputs is computed as:

---

[26] As discussed in Sect. 3.4, Slaney (1998) covered the frequency range [133, 6855] Hz with a filter-bank of 40 filters. In the comparative performance evaluations of multiple speech features, presented in Sects. 5–7, we will conform to the frequency range of Slaney (1998), and will use a LFCC filter-bank of 40 filters, referred to as LFCC-FB40.

$$S_i = \log_{10}\left(\sum_{k=0}^{N-1} |S(k)|^2 \cdot H_i(k)\right), \quad i = 1, 2, ..., M, \tag{3.4}$$

where S_i is the output of the ith filter, $|S(k)|^2$ is the power spectrum, and N is the DFT size.

Finally, the LFCC are obtained after performing decorrelation of the filter-bank outputs via the DCT:

$$LFCC(r) = \sum_{i=1}^{M} S_i \cos\left(\frac{r(i - 0.5)\pi}{B}\right), \quad r = 0, 1, ..., R - 1. \tag{3.5}$$

Here r is the LFCC index, and $R \leq M$ is the total number of unique LFCC that can be computed. For larger R, the values of the LFCC with index $r \geq M$ mirror those of the first M coefficients. In all further discussions and in the experimental evaluation in Sects. 5–7, we consider the filter-bank version of the LFCC.

3.2 The MFCC-FB20

In brief, the origins of the MFCC can be traced back to the work of Bridle and Brown (1974), who used a set of 19 cepstral coefficients obtained after applying the cosine transform on the logarithmically compressed outputs of a filter-bank with 19 nonuniformly spaced band-pass filters. In fact, Bridle and Brown do not explain[27] the reasoning for this particular design of the filter-bank, but describe it as "a typical arrangement of variation of channel spacing with centre frequency," which covers the frequency range [180, 4000] Hz. They acknowledge that this filter-bank already existed as front-end processing of their computer and was used for the needs of speech coding (in the analysis part of a vocoder). However, Bridle and Brown (1974), did not make use of the filter-bank output directly but applied the cosine transform to approximate the principal component analysis used in Klein et al. (1970) on the task of vowel identification and in Pols (1971) on the task of small vocabulary isolated word recognition. It was already known (King 1971), that the principal component analysis and cosine transform are essentially equivalent with respect to their decorrelation effect.

In addition, Bridle and Brown (1974), instead of discarding the cepstral coefficients which contribute less to the recognition performance, introduced weighting according to their relative importance. The weighs were discrete numbers, and the highest went to the cepstral coefficients with indexes one and two, the next two were for these with indexes three and four, and the next two for these with indexes five and six.

---

[27] However, Bridle and Brown (1974), referred to an earlier work of H. Yilmaz (1967) A theory of speech perception. *Bulletin of Mathematical Biophysics* 29(4):739–825, where spectrum-shape method is used for normalization of the speaker characteristics and spectral colorization.

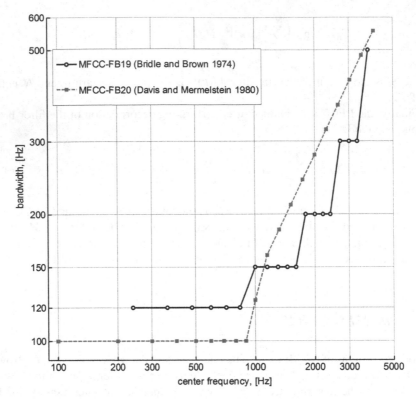

Fig. 3.2 Comparison between the nonlinear filter-bank used by Bridle and Brown (1974) – *solid line with markers* "o" – and the filter-bank used by Davis and Mermelstein (1980) – *dashed line with marker* "■"

The coefficient with index zero and these with indexes bigger than six received the lowest weight factor.

Later on, in the work of Davis and Mermelstein (1980), the MFCC are described as a set of discrete cosine transform-decorrelated parameters, which are computed through a transformation of the logarithmically compressed filter-output energies, derived through a perceptually spaced filter-bank of triangular filters that is applied on the discrete Fourier transform (DFT)-ed speech signal.

As illustrated in Figs. 3.2 and 3.3, the filter-bank used by Davis and Mermelstein (1980), is comprised of 20 equal-height filters, spaced according the Koenig scale (1.2), and covering the frequency range [0, 5000] Hz. Thus, in the following, we refer to this implementation as to MFCC-FB20. The center frequencies of the first ten filters, residing in the frequency range [100, 1000] Hz, are linearly spaced, and the next ten have center frequencies logarithmically spaced between 1000 and 4000 Hz. The choice of center frequency f_{c_i} for the ith filter can be approximated (Skowronski 2004) as:

$$f_{c_i} = \begin{cases} 100 \cdot i, & i = 1, ..., 10 \\ f_{c_{10}} \cdot 2^{0.2(i-10)}, & i = 11, ..., 20 \end{cases}, \tag{3.6}$$

where the center frequency f_{c_i} is assumed in Hz.

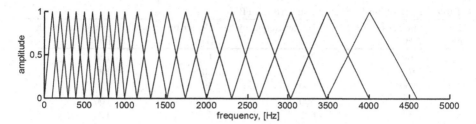

Fig. 3.3 Mel-spaced filter-bank of equal-height filter according to Davis and Mermelstein (1980). The center frequencies of the first ten filters are linearly spaced, and next ten have logarithmically spaced center frequencies

As illustrated in Fig. 3.3, the endpoints of each triangular filter are determined by the center frequencies of adjacent filters, and therefore, the bandwidth of these filters is not an independent variable. More precisely, the bandwidths of the filters are determined by the spacing between the center frequencies of the adjacent filters, which on its hand is a function of the sampling rate of the signal and the number of the filters in the filter-bank. Therefore, for a given sampling frequency, increase of the number of filters results in decrease of their bandwidth. Table 3.1 summarizes the central frequencies and the bandwidth of the filters in the filter-bank of MFCC-FB20.

As it becomes clear from the discussion in Sect. 1.4, the MFCC filter-banks as they are computed in Bridle and Brown (1974) and Davis and Mermelstein (1980) include approximations of the critical bandwidths that do not completely conform to today's understanding of the critical band concept. Nevertheless, the introduction of perceptually motivated filter-banks in the computation of the cepstral coefficients had a great impact on the evolution of the speech parameterization methods and contributed to improving the performance of the speech recognition technology.

In the remainder of this section, we recapitulate the computation of the MFCC-FB20 partially following the exposition in Huang et al. (2001). In brief, let us denote with n the discrete-time index, and with $x(n)$ a discrete-time speech signal that has been sampled with sampling frequency f_s. Let us consider that the signal $x(n)$ has been pre-processed as explained in Sect. 2.2 and has been segmented in frames with length of N samples. Each speech segment obtained to this end, represented by $s(n)$, $n = 0, 1, ..., N - 1$, which was pre-emphasized and weighted by the Hamming window, is subject to the DFT,

$$S(k) = \sum_{n=0}^{N-1} s(n) \cdot \exp\left(\frac{-j2\pi nk}{N}\right), \quad k = 0, 1, ..., N - 1, \quad (3.7)$$

where k is the index of the Fourier coefficients, $S(k)$.

Table 3.1 The filter-bank used in the MFCC-FB20

Filter no.	Lower frequency [Hz]	Higher frequency [Hz]	Center frequency [Hz]	Filter bandwidth [Hz]
1	0	200	100	100
2	100	300	200	100
3	200	400	300	100
4	300	500	400	100
5	400	600	500	100
6	500	700	600	100
7	600	800	700	100
8	700	900	800	100
9	800	1000	900	100
10	900	1149	1000	125
11	1000	1320	1149	160
12	1149	1516	1320	184
13	1320	1741	1516	211
14	1516	2000	1741	242
15	1741	2297	2000	278
16	2000	2639	2297	320
17	2297	3031	2639	367
18	2639	3482	3031	422
19	3031	4000	3482	485
20[a]	3482	4595	4000	557
21[a]	4000	5278	4595	639
22[a]	4595	6063	5278	734
23[a]	5278	6964	6063	843
24[a]	6063	8000	6964	969

[a] These filters are present only in the wideband version ([0, 8000] Hz) of the filter-bank

Next, a filter-bank $H_i(k)$, $i = 1, 2, ..., M$, with M equal-height triangular filters is constructed. Each of these M filters is defined as:

$$H_i(k) = \begin{cases} 0 & \text{for} \quad k < f_{b_{i-1}} \\ \dfrac{(k - f_{b_{i-1}})}{(f_{b_i} - f_{b_{i-1}})} & \text{for} \quad f_{b_{i-1}} \le k \le f_{b_i} \\ \dfrac{(f_{b_{i+1}} - k)}{(f_{b_{i+1}} - f_{b_i})} & \text{for} \quad f_{b_i} \le k \le f_{b_{i+1}} \\ 0 & \text{for} \quad k > f_{b_{i+1}} \end{cases}, \quad i = 1, 2, ..., M, \qquad (3.8)$$

where the index i stands for the ith filter, f_{b_i} are the boundary points of the filters, and $k = 0, 1, ..., N - 1$ corresponds to the kth coefficient of the N-point DFT. The boundary points f_{b_i} are expressed in terms of position. Their relative position depends on the sampling frequency f_s and the number of points N in the DFT, and they are computed as:

$$f_{b_i} = \left(\frac{N}{f_s}\right) \cdot \hat{f}_{mel}^{-1}\left(\hat{f}_{mel}(f_{low}) + i \cdot \frac{\hat{f}_{mel}(f_{high}) - \hat{f}_{mel}(f_{low})}{M+1}\right). \tag{3.9}$$

Here, the function $\hat{f}_{mel}(.)$ stands for the transformation,

$$\hat{f}_{mel} = 1127 \cdot \ln\left(1 + \frac{f_{lin}}{700}\right), \tag{3.10}$$

the f_{low} and f_{high} are respectively the low and the high boundary frequency for the entire filter-bank, M is the number of filters, and $\hat{f}_{mel}^{-1}$ is the inverse to (3.10) transformation, formulated as:

$$\hat{f}_{mel}^{-1} = 700 \cdot \left[\exp\left(\frac{\hat{f}_{mel}}{1127}\right) - 1\right]. \tag{3.11}$$

Here, the sampling frequency f_s and the frequencies f_{low}, f_{high}, and f_{lin} are in Hz, and the $\hat{f}_{mel}$ is in Mels. Equation 3.9 provides that the boundary points of the filters are uniformly spaced in the Mel scale.

Finally, according to Bridle and Brown (1974), the MFCC speech features are obtained after applying the DCT, as:

$$MFCC(r) = \sum_{i=1}^{M} S_i \cdot \cos\left(\frac{r(i - 0.5)\pi}{M}\right), r = 0, 1, ..., R - 1, \tag{3.12}$$

where M is the number of filters in the filter-bank, $R \leq M$ is the number of unique cepstral coefficients which can be computed. For larger R, the values of the MFCC with index $r \geq M$ mirror these of the first M coefficients. Here, S_i is formulated as the "log-energy output of the ith filter" and is understood as:

$$S_i = \log_{10}\left(\sum_{k=0}^{N-1} |S(k)| \cdot H_i(k)\right), \quad i = 1, 2, ..., M. \tag{3.13}$$

In this case, the log-energy output S_i of each filter is derived through the magnitude spectrum $|S(k)|$, and the filter-bank $H_i(k)$ defined in (3.8). It has to be mentioned here that since S_i is derived through the magnitude spectrum $|S(k)|$, and not through the power spectrum $|S(k)|^2$, it does not comply with the Parseval's definition of energy as sum of squared terms.

In the work of Davis and Mermelstein (1980), Eq. 3.12 appears as in Bridle and Brown (1974); however, other authors, for instance (Huang et al. 2001), consider the range of summation as $i = 0, 1, ..., M - 1$, that is, Eq. 3.12 is given as:

$$MFCC(r) = \sum_{i=0}^{M-1} S_i \cdot \cos\left(\frac{r(i - 0.5)\pi}{M}\right), \quad r = 0, 1, ..., R - 1. \tag{3.14}$$

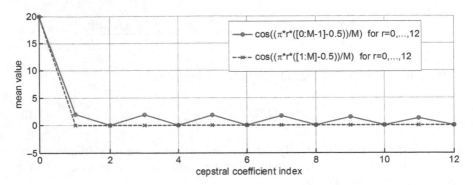

Fig. 3.4 *Dashed line with marker* "x" stands for Eq. 3.12 and *solid line with marker* "o" stands for Eq. 3.14

However, Eq. 3.12 seems to be more advantageous when compared to (3.14), because it guarantees zero-mean value for the cepstral coefficients for any $0<r<R$, as there is no misbalance between the "positive" and "negative" terms of cos(.). On the other hand, for Eq. 3.14, zero-mean value is guaranteed only for the coefficients with even index $r = \{2, 4, 6, ...\}$.

Figure 3.4 illustrates the effect on the values of the cos(.) function for the different implementations: Eq. 3.12 with dashed line and marker "x," and Eq. 3.14 with solid line and marker "o."

Furthermore, when in Eq. 3.14 the term $(i - 0.5)$ is replaced with $(i + 0.5)$, as in (3.15), the balance between the negative and positive portions of the cos(.) function is recovered and the resultant MFCC are guaranteed to have zero-mean value, given some balanced speech signal.

$$MFCC(r) = \sum_{i=0}^{M-1} S_i \cdot \cos\left(\frac{r(i + 0.5)\pi}{M}\right), \quad r = 0, 1, ..., R - 1. \qquad (3.15)$$

Even though it is well known that the triangular shape of the filters only roughly approximates the shape of the auditory filters of the human auditory system, and that the known relationship (1.7) between center frequency of the filter and critical bandwidth is not used, the general concept of the MFCC paradigm led to a significant advance in the speech parameterization research. A number of researchers elaborated on the original MFCC design, and novel, biologically motivated speech parameterizations emerged. In the following sections, we discuss two popular MFCC implementations, known as the HTK MFCC (Young et al. 1995) and the Auditory Toolbox (Slaney 1998) MFCC. The latter ones are also the default speech features in the open-source CMU Sphinx speech recognizer.

3.3   The HTK MFCC-FB24

Another implementation of the MFCC that is now widely used in the speech processing community was created within the framework of the Cambridge hidden Markov models (HMM) Toolkit (HTK) described in Young et al. (1995). Here, this implementation is referred to as HTK MFCC-FB24. The designation HTK MFCC-FB24 reflects the number of filters $(M = 24)$ recommended by Young (1996), for signal bandwidth of 8000 Hz. In the MFCC implementation of HTK, similar to the approach presented in Davis and Mermelstein (1980), a filter-bank of equal-height filters is assumed.

In detail, the MFCC-FB24 speech parameterization makes use of the definition (1.5) of the Mel frequency, which is rewritten here for convenience as:

$$\hat{f}_{mel} = 2595 \cdot \log_{10}\left(1 + \frac{f_{lin}}{700}\right). \tag{3.16}$$

Equation 3.16 relates the linear frequency f_{lin} in Hertz to the warped frequency $\hat{f}_{mel}$ in Mels.

In the HTK implementation of the MFCC, the limits of the frequency range are the parameters that define the basis for the filter-bank design. More specifically, specifying the lower and the higher boundaries of the frequency range of the entire filter-bank, $\hat{f}_{low}$ and $\hat{f}_{high}$, respectively, is considered as the starting point for the computation of the filter bandwidth,

$$\Delta\hat{f} = \frac{\hat{f}_{high} - \hat{f}_{low}}{M + 1}, \tag{3.17}$$

which serves as a footstep in the definition of the center frequencies of the individual filters. The center frequency $\hat{f}_{c_i}$ of the ith filter in Mels is given by:

$$\hat{f}_{c_i} = \hat{f}_{low} + i \cdot \Delta\hat{f}, \quad i = 1, 2, ..., M - 1, \tag{3.18}$$

where M is the total number of filters in the filter-bank. The center frequencies of the filters are transformed to Hz, as:

$$f_{c_i} = 700 \cdot \left(10^{\hat{f}_{c_i}/2595} - 1\right). \tag{3.19}$$

Next, the shape of the individual triangular filters is defined as

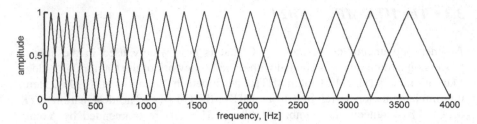

Fig. 3.5 Mel-spaced filter-bank composed of 20 equal-height filters with logarithmically spaced center frequencies

$$
H_i(k) = \begin{cases}
0 & \text{for} \quad k < f_{b_{i-1}} \\[2ex]
\dfrac{(k - f_{b_{i-1}})}{(f_{b_i} - f_{b_{i-1}})} & \text{for} \quad f_{b_{i-1}} \leq k \leq f_{b_i} \\[2ex]
\dfrac{(f_{b_{i+1}} - k)}{(f_{b_{i+1}} - f_{b_i})} & \text{for} \quad f_{b_i} \leq k \leq f_{b_{i+1}} \\[2ex]
0 & \text{for} \quad k > f_{b_{i+1}}
\end{cases} \quad , \quad i = 1, 2, ..., M, \qquad (3.20)
$$

where the index i stands for the ith filter $H_i(.)$, f_{b_i} are the boundary points of the filters, and $k = 0, 1, ..., N - 1$ corresponds to the kth coefficient of the N-point DFT. In fact, the boundary points, f_{b_i}, of each filter are the center frequencies, f_{c_i}, of its left and right neighbors.

In Young (1996), 24 filters were suggested for speech bandwidth of 8000 Hz. Here, we also discuss a narrowband version of the HTK MFCC-FB24 filter-bank for speech bandwidth of 4000 Hz and sampling frequency of 8000 Hz, which we refer to as HTK MFCC-FB20 (illustrated in Fig. 3.5). However, there is yet another interpretation of the HTK filter-bank, which uses 24 filters for the frequency range [0, 4000] Hz as in Skowronski (2004) and Benesty et al. (2008). Table 3.2 presents the 24-filter filter-bank used in the narrowband version of the HTK MFCC-FB24. The 20-filter filter-bank used in the HTK MFCC-FB20, and yet another version of the HTK filter-bank, with 26 filters for the frequency range [0, 8000] Hz, which is the default filter-bank in the recent versions of the HTK (Young et al. 2006) are shown in Appendix I.

The computation of the HTK MFCC parameters can be summarized as follows. Let us denote with n the discrete-time index, and with $x(n)$ a discrete-time speech signal that has been sampled with sampling frequency f_s. Let us consider that the signal $x(n)$ has been pre-processed as explained in Sect. 2.2, and has been segmented in frames with length of N samples. Each speech segment obtained to this end, represented by $s(n)$, $n = 0, 1, ..., N - 1$, which was pre-emphasized and weighted by the Hamming window, is subject to the DFT,

$$
S(k) = \sum_{n=0}^{N-1} s(n) \cdot \exp\left(\frac{-j2\pi nk}{N}\right), \quad k = 0, 1, ..., N - 1. \qquad (3.21)
$$

Table 3.2 The HTK MFCC-FB24 with a filter-bank of 24 filters in the frequency range [0, 4000] Hz

Filter no.	Lower frequency [Hz]	Higher frequency [Hz]	Center frequency [Hz]	Filter bandwidth [Hz]
1	0	115	55	58
2	55	180	115	63
3	115	249	180	67
4	180	324	249	72
5	249	406	324	79
6	324	493	406	85
7	406	587	493	91
8	493	689	587	98
9	587	799	689	106
10	689	918	799	115
11	799	1046	918	124
12	918	1184	1046	133
13	1046	1333	1184	144
14	1184	1494	1333	155
15	1333	1668	1494	168
16	1494	1855	1668	181
17	1668	2058	1855	195
18	1855	2276	2058	211
19	2058	2511	2276	227
20	2276	2766	2511	245
21	2511	3040	2766	265
22	2766	3336	3040	285
23	3040	3655	3336	308
24	3336	4000	3655	332

In Eq. 3.21 n is the index of the time-domain samples, and k is the index of the Fourier coefficients $S(k)$. Next, $S(k)$ is used for computing the power spectrum $|S(k)|^2$, which then acts as input for the filter-bank $H_i(.)$ defined in (3.20). At the next step, the filter-bank output is logarithmically compressed as

$$S_i = \ln\left(\sum_{k=0}^{N-1} |S(k)|^2 \cdot H_i(k)\right), \quad i = 1, 2, ..., M \tag{3.22}$$

and then the filter-bank outputs are decorrelated by the DCT, to provide the HTK MFCC speech features:

$$MFCC_{HTK}(r) = \sqrt{\frac{2}{M}} \sum_{i=1}^{M} S_i \cdot \cos\left(\frac{r(i-0.5)\pi}{M}\right), r = 0, 1, ..., R - 1. \tag{3.23}$$

Here, M is the number of filters in the filter-bank and $R \leq M$ is the number of unique cepstral coefficients which can be computed. For larger R, the values of the

MFCC with index $r \geq M$ mirror these of the first M coefficients. The scale factor $\sqrt{2/M}$ is for making the DCT matrix orthogonal. In addition, the cepstral coefficient with index $r = 0$ is multiplied by the term $1/\sqrt{2}$ for the same reason. The number of cepstral coefficients, R, which are computed, is an application-dependent issue. For instance, on the monophone recognition and the speech recognition tasks, the default settings of HTK (Young et al. 2006) consider the first thirteen or when the one with index zero is excluded, the first twelve cepstral coefficients. On the speaker recognition task (Fauve et al. 2007) most often researchers make use of the first nineteen cepstral coefficients, where the one with index zero is excluded.

3.4   The MFCC-FB40

In the Auditory Toolbox (Slaney 1998) the MFCC are computed through a filter-bank of 40 filters, and therefore in the following we refer to this implementation as MFCC-FB40. In brief, assuming sampling frequency $f_s = 16000$ Hz, Slaney implemented a filter-bank of 40 equal-area filters, which cover the frequency range [133.3, 6855] Hz. The center frequencies, f_{c_i}, of the first 13 of them are linearly spaced, $N_{lin} = 13$, in the range [200, 1000] Hz with step of ~66.7 Hz and the ones of the next 27 are logarithmically spaced, $N_{log} = 27$, in the range [1071, 6400] Hz, as:

$$f_{c_i} = \begin{cases} 133.33333 + 66.66667 \cdot i, & i = 1, 2, ..., N_{lin} \\ f_{N_{lin}} \cdot F_{\log}^{i - N_{lin}}, & i = N_{lin} + 1, N_{lin} + 2, ..., N_{total} \end{cases} \quad (3.24)$$

Here, the log-factor $F_{\log} = \exp\left(\ln(f_{c_{tottal}}/1000)/N_{\log}\right)$ is $F_{\log} = 1.07117029$, when the total number of the logarithmically spaced filters $N_{\log} = 27$, and the desired center frequency for the last of them, $N_{tottal} = N_{lin} + N_{\log} = 40$, is $f_{c_{40}} = 6400$ Hz. Table 3.3 shows the MFCC-FB40 filter-bank design computed as in Slaney (1998). On the analogy of the equal-height filters (3.20), each of the equal-area filters is defined as:

$$H_i(k) = \begin{cases} 0 & \text{for } k < f_{b_{i-1}} \\ \dfrac{2(k - f_{b_{i-1}})}{(f_{b_i} - f_{b_{i-1}})(f_{b_{i+1}} - f_{b_{i-1}})} & \text{for } f_{b_{i-1}} \leq k \leq f_{b_i} \\ \dfrac{2(f_{b_{i+1}} - k)}{(f_{b_{i+1}} - f_{b_i})(f_{b_{i+1}} - f_{b_{i-1}})} & \text{for } f_{b_i} \leq k \leq f_{b_{i+1}} \\ 0 & \text{for } k > f_{b_{i+1}} \end{cases} , \quad i = 1, 2, ..., M, \quad (3.25)$$

where i stands for the ith filter, f_{b_i} are $M + 2$ boundary points that specify the M filters, and $k = 1, 2, ..., N$ corresponds to the kth coefficient of the N-point DFT.

Table 3.3 The filter-bank used in the MFCC-FB40 as in Slaney (1998)

Filter no.	Lower frequency [Hz]	Higher frequency [Hz]	Center frequency [Hz]	Filter bandwidth [Hz]
1	133	267	200	67
2	200	333	267	67
3	267	400	333	67
4	333	467	400	67
5	400	533	467	67
6	467	600	533	67
7	533	667	600	67
8	600	733	667	67
9	667	800	733	67
10	733	867	800	67
11	800	933	867	67
12	867	1000	933	67
13	933	1071	1000	69
14	1000	1147	1071	74
15	1071	1229	1147	79
16	1147	1317	1229	85
17	1229	1410	1317	91
18	1317	1511	1410	97
19	1410	1618	1511	104
20	1511	1733	1618	111
21	1618	1857	1733	120
22	1733	1989	1857	128
23	1857	2130	1989	137
24	1989	2282	2130	147
25	2130	2444	2282	157
26	2282	2618	2444	168
27	2444	2805	2618	181
28	2618	3004	2805	193
29	2805	3218	3004	207
30	3004	3447	3218	222
31	3218	3692	3447	237
32	3447	3955	3692	254
33[a]	3692	4237	3955	273
34[a]	3955	4538	4237	292
35[a]	4237	4861	4538	312
36[a]	4538	5207	4861	335
37[a]	4861	5578	5207	359
38[a]	5207	5975	5578	384
39[a]	5578	6400	5975	411
40[a]	5975	6855	6400	440

[a] These filters are not present in the narrowband version ([0, 4000] Hz) of Slaney's MFCC

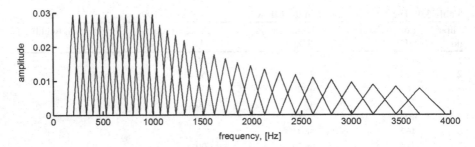

Fig. 3.6 Mel-spaced filter-bank composed of equal area filters for the range [0, 4000] Hz. The center frequencies of the first 13 filters are linearly spaced and the ones of the next 19 are logarithmically spaced

The boundary points f_{b_i} are expressed in terms of position, as specified above. The key to equalization of the area below the filters (3.25) lies in the term:

$$\frac{2}{\left(f_{b_{i+1}} - f_{b_{i-1}}\right)}. \tag{3.26}$$

Due to the term (3.26), the filter-bank (3.25) is normalized in such a way that the sum of coefficients for every filter equals one. Thus, the ith filter satisfies:

$$\sum_{k=1}^{N} H_i(k) = 1, \quad \text{for} \quad i = 1, 2, ..., M. \tag{3.27}$$

In setups where sampling frequency of $f_s = 8000$ Hz is considered, only the filters placed below 4000 Hz are retained. Figure 3.6 illustrates the first 32 filters, which cover the frequency range [133.3, 3955] Hz and have spacing and bandwidth according to the original scheme of Slaney (1998), described above. As shown in the figure, the center frequencies of the first 13 filters are linearly spaced in the range [200, 1000] Hz and the ones of the next 19 are logarithmically spaced (3.24) in the range [1071, 3692] Hz. For clarity of exposition, in the following, this narrowband version of Slaney's MFCC that corresponds to sampling rate of 8000 Hz and signal bandwidth of 4000 Hz is referred to as MFCC-FB32.

The computation of the MFCC-FB40 can be summarized as follows. Let us denote with n the discrete-time index, and with $x(n)$ a discrete-time speech signal that has been sampled with sampling frequency f_s. Let us consider that the signal $x(n)$ has been pre-processed as explained in Sect. 2.2 and has been segmented in frames with length of N samples. Each speech segment obtained to this end, represented by $s(n)$, $n = 0, 1, ..., N - 1$, which was pre-emphasized and weighted by the Hamming window, is subject to the DFT,

$$S(k) = \sum_{n=0}^{N-1} s(n) \cdot \exp\left(\frac{-j2\pi nk}{N}\right), \quad k = 0, 1, ..., N - 1. \tag{3.28}$$

Here, n is the index of the time-domain samples, and k is the index of the Fourier coefficients $S(k)$. Next, $S(k)$ is used for computing the amplitude spectrum $|S(k)|$, and then the equal-area filter-bank $H_i(.)$ (3.25) is employed in the computation of the log-energy output:

$$S_i = \log_{10}\left(\sum_{k=0}^{N-1} |S(k)| \cdot H_i(k)\right), \quad i = 1, 2, ..., M. \tag{3.29}$$

Finally, according to the implementation in Slaney (1998), the discrete cosine transform of type DCT-II (3.30) is applied to obtain the MFCC-FB40 cepstral coefficients. Adhering precisely to the DCT-II formulation, Slaney assumed the indexes of the filter-bank output S_i as $i = 0, 1, ..., M-1$, so here we shall use S_{i+1}:

$$MFCC(r) = \sqrt{\frac{2}{M}} \sum_{i=0}^{M-1} S_{i+1} \cdot \cos\left(\frac{r(i+0.5)\pi}{M}\right), r = 0, 1, ..., R-1. \tag{3.30}$$

Here M is the number of filters in the filter-bank, and $R \leq M$ is the number of unique cepstral coefficients which can be computed. For larger R, the values of the MFCC with index $r \geq M$ mirror these of the first M coefficients. The scale factor $\sqrt{2/M}$ makes the DCT matrix orthogonal, and for the same reason the cepstral coefficient with index $r = 0$ is next multiplied by the term $\sqrt{2}/2$. The number of unique cepstral coefficients is $R \leq 32$ or $R \leq 40$ depending on the filter-bank design (MFCC-FB32 or MFCC-FB40, respectively), but the actual value of R is often smaller and depends on the specific application.

3.5 The HFCC-E-FB29

The Human Factor Cepstral Coefficients (HFCC), introduced by Skowronski and Harris (2004), represent the most recent update of the MFCC filter-bank. Similar to the other MFCC implementations discussed in Sects. 3.2–3.4, the HFCC do not proclaim to be a perceptual model of the human auditory system, but rather is a biologically inspired feature extraction scheme.

Assuming sampling frequency of 12500 Hz, Skowronski and Harris (2004) proposed the HFCC filter-bank composed of 29 Mel-warped equal-height filters, which cover the frequency range [0, 6250] Hz. In Fig. 3.7, only the first 24 filters, which cover the frequency range of [0, 4000] Hz, are shown. As illustrated in the figure, in the HFCC scheme, the overlapping among the filters is different from the traditional – and one filter can overlap not only with its closest neighbors but also with more remote neighbors.

The most significant difference in the HFCC, when compared to the earlier MFCC schemes, is that the filter bandwidth is decoupled from the filter spacing.

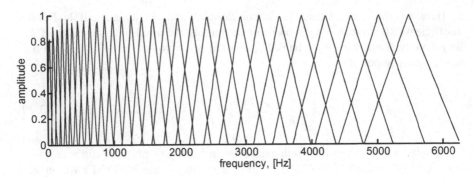

Fig. 3.7 The HFCC filter-bank proposed by Skowronski and Harris (2004)

More specifically, the filter bandwidth in the HFCC is derived from the equivalent rectangular bandwidth (ERB) approximation of the critical bandwidth, introduced by Moore and Glasberg (1983):

$$ERB = 6.23 \frac{f_c^2}{10^6} + 93.39 \frac{f_c}{10^3} + 28.52, \tag{3.31}$$

where f_c is the center frequency of the individual filters in Hz. The filter bandwidth computed by (3.31) is further scaled by a constant, which Skowronski and Harris referred to as E-factor. One consequence from the decoupling of the central frequency and the filter bandwidth is that the latter becomes an independent variable, which can be adjusted (through the E-factor) in a data-dependent manner or based on prior knowledge about the target application. This can be advantageous for fine-tuning of the system accuracy and robustness in diverse environmental conditions.

In brief, the HFCC filter-bank design, as it is described in Skowronski (2004), consists in the following steps: First, the low f_{low} and high f_{high} boundaries of the entire filter-bank and the number M of filters are specified. The center frequencies f_{c_1} and f_{c_M} of the first and the last of the filters, respectively, are computed as:

$$f_{c_i} = \frac{1}{2}\left(-\bar{b} + \sqrt{\bar{b}^2 - 4\bar{c}}\right), \tag{3.32}$$

where the index i is either 1 or M, and the coefficients $\bar{b}$ and $\bar{c}$, defined as:

$$\bar{b} = \frac{b - \hat{b}}{a - \hat{a}} \text{ and } \bar{c} = \frac{c - \hat{c}}{a - \hat{a}}, \tag{3.33}$$

receive different values for the two cases. The values of the constants a, b, c come from (3.31) and are $6.23 \cdot 10^{-6}, 93.39 \cdot 10^{-3}, 28.52$, respectively. For the first filter, the values of the coefficients $\hat{a}, \hat{b}, \hat{c}$ are computed as:

$$\hat{a} = \frac{1}{2} \cdot \frac{1}{700 + f_{low}}, \quad \hat{b} = \frac{700}{700 + f_{low}}, \quad \hat{c} = -\frac{f_{low}}{2} \cdot \left(1 + \frac{700}{700 + f_{low}}\right). \tag{3.34}$$

For the last filter, these coefficients are:

$$\hat{a} = -\frac{1}{2} \cdot \frac{1}{700 + f_{high}}, \quad \hat{b} = -\frac{700}{700 + f_{high}}, \quad \hat{c} = \frac{f_{high}}{2} \cdot \left(1 + \frac{700}{700 + f_{high}}\right). \quad (3.35)$$

Once the center frequencies of the first and the last filters are computed, the center frequencies of the filters lying between them are easily computed, since they are equidistant on the Mel scale. The step $\Delta\hat{f}$ between the center frequencies of adjacent filters is computed as:

$$\Delta\hat{f} = \frac{\hat{f}_{c_M} - \hat{f}_{c_1}}{M - 1}, \quad (3.36)$$

where all frequencies are in Mels. The conversions from Hertz to Mel, $f_{c_1} \to \hat{f}_{c_1}$ and $f_{c_M} \to \hat{f}_{c_M}$, are given as:

$$\hat{f}_{c_i} = 2595 \cdot \log_{10}\left(1 + \frac{f_{c_i}}{700}\right). \quad (3.37)$$

Then, the center frequencies $\hat{f}_{c_i}$ are computed as:

$$\hat{f}_{c_i} = \hat{f}_{c_1} + (i - 1) \cdot \Delta\hat{f}, \text{ for } i = 2, ..., M - 1. \quad (3.38)$$

Next, through (3.39), the reverse transformation from Mel to Hertz, $\hat{f}_{c_i} \to f_{c_i}$, is performed as,

$$f_{c_i} = 700 \cdot \left(10^{\hat{f}_{c_i}/2595} - 1\right), \quad (3.39)$$

and through (3.31) the ERB_i for each center frequency f_{c_i} is computed. Finally, the low and high frequencies f_{low_i} and f_{high_i}, respectively, of the ith filter are derived through the constraints:

$$ERB_i = \frac{1}{2} \cdot \left(f_{high_i} - f_{low_i}\right) \quad (3.40)$$

$$\hat{f}_{c_i} = \frac{1}{2} \cdot \left(\hat{f}_{high_i} - \hat{f}_{low_i}\right), \quad (3.41)$$

which is transformed to:

$$f_{low_i} = -(700 + ERB_i) + \sqrt{(700 + ERB_i)^2 + f_{c_i}(f_{c_i} + 1400)}, \quad (3.42)$$

$$f_{high_i} = f_{low_i} + 2 \cdot ERB_i. \quad (3.43)$$

Table 3.4 The HFCC-FB29 filter-bank extrapolated for the frequency range [0, 8000] Hz

Filter no.	Lower frequency [Hz]	Higher frequency [Hz]	Center frequency [Hz]	Filter bandwidth [Hz]
1	12	61	31	31
2	61	122	89	37
3	110	183	151	43
4	171	269	219	49
5	244	342	291	56
6	317	427	370	64
7	391	525	455	72
8	476	623	546	81
9	562	732	645	91
10	659	854	751	102
11	757	977	866	114
12	879	1111	991	127
13	989	1270	1124	141
14	1123	1428	1269	157
15	1270	1599	1425	174
16	1416	1794	1593	193
17	1575	1990	1775	214
18	1746	2209	1971	237
19	1941	2454	2183	262
20	2136	2710	2411	290
21	2356	2991	2658	321
22	2588	3296	2924	355
23	2844	3613	3211	393
24	3113	3967	3521	435
25	3406	4358	3855	481
26	3723	4773	4216	533
27	4053	5225	4606	591
28	4419	5713	5026	655
29	4797	6238	5480	727
30[a]	5181	6737	5970	807
31[a]	5595	7343	6497	896
32[a]	6043	7931	7064	994

[a] These filters are present only for wideband signal with bandwidth [0, 8000] Hz

With all parameters computed through the procedure described with (3.32) ÷ (3.43), the design of the HFCC filter-bank is completed. Table 3.4 shows the HFCC-FB29, extrapolated for the frequency range [0, 8000] Hz. In Sects. 5 and 7, we also consider three additional designs of the HFCC filter-bank, with $M = 19$, 23 and 29 filters and evaluate them for various values of the E-factor.

In brief, the computation of the HFCC-FB29 speech parameters is summarized as follows. Let us denote with n the discrete-time index, and with $x(n)$ a discrete-time speech signal that has been sampled with sampling frequency f_s. Let us consider that the signal $x(n)$ has been pre-processed as explained in Sect. 2.2 and has been segmented in frames with length of N samples. Each speech segment obtained to this end, represented by $s(n)$, $n = 0, 1, ..., N - 1$, which was pre-emphasized and weighted by the Hamming window, is subject to the DFT,

$$S(k) = \sum_{n=0}^{N-1} s(n) \cdot \exp\left(\frac{-j2\pi nk}{N}\right), k = 0, 1, ..., N-1. \qquad (3.44)$$

Here, n is the index of the time-domain samples, and k is the index of the Fourier coefficients $S(k)$. Next, $S(k)$ is used for computing the magnitude spectrum $|S(k)|$, and then similarly to the MFCC scheme of Davis and Mermelstein (1980), the log-energy filter-bank outputs are computed as

$$S_i = \log_{10}\left(\sum_{k=0}^{N-1} |S(k)| \cdot H_i(k)\right), \quad i = 1, 2, ..., M. \qquad (3.45)$$

Finally, the DCT (3.46) in the DCT-II form is applied to decorrelate the HFCC speech features:

$$HFCC(r) = \sqrt{\frac{2}{M}} \sum_{i=0}^{M-1} S_{i+1} \cdot \cos\left(\frac{r(i+0.5)\pi}{M}\right), \quad r = 0, 1, ..., R-1. \qquad (3.46)$$

Here, M is the number of filters in the filter-bank and $R \leq M$ is the number of unique cepstral coefficients which can be computed. For larger R, the values of the HFCC with index $r \geq M$ mirror these of the first M coefficients. As in the other speech parameterization methods, the actual number of cepstral coefficients, R, to be computed is an application-dependent issue. The scale factor $\sqrt{2/M}$ is used for making the DCT matrix orthogonal. Here, it is also important to note that in the original implementation[28] of the HFCC, the cepstral coefficient with index $r = 0$ is excluded from the final speech feature vector, and is replaced by the energy of the speech frame, computed directly from the original signal, before the pre-emphasis and the other pre-processing steps.

3.6   The PLP-FB19

Aiming to surmount the shortcomings of the linear predictive (LP) analysis, Hermansky (1990) proposed the now venerable perceptual linear predictive (PLP) analysis of speech. This speech parameterization technique takes advantage of the latest advances in psychoacoustics known at that time and incorporates three engineering approximations of the properties of human hearing, among which are: (i) the critical-band spectral resolution, (ii) the equal-loudness curve, and (iii) the intensity-loudness power law. In the following, we outline the PLP speech parameterizations, following closely the exposition in Hermansky (1990).

---

[28] HFCC code: http://www.cnel.ufl.edu/~markskow/

Let us denote with n the discrete-time index, and with $x(n)$, $n = 0, 1, ..., N - 1$ a discrete-time speech signal with length of N samples that has been sampled with sampling frequency f_s. Let us consider that the signal $x(n)$ has been pre-processed accordingly as explained in Sect. 2.2, except the pre-emphasis step, which in the PLP parameterization scheme is not part of the pre-processing of speech but is performed in the frequency domain. The latter is equivalent to setting the pre-emphasis filter coefficient $a = 0$ in Eq. 2.6. Then, applying N-point DFT on the discrete-time input signal $s(n)$, we obtain the short-time spectrum $S(k)$ as:

$$S(k) = \sum_{n=0}^{N-1} s(n) \cdot \exp\left(\frac{-j2\pi nk}{N}\right), k = 0, 1, ..., N - 1, \qquad (3.47)$$

where k is the discrete frequency index. Next, the power spectrum $|S(k)|^2$ is warped along the frequency axis, which results in $|S(f_{Bark})|^2$, by making use of the approximate transformation between the linear frequency in Hertz and the Bark frequency,

$$f_{Bark} = 6 \ln\left(\frac{f_{lin}}{600} + \sqrt{1 + \left(\frac{f_{lin}}{600}\right)^2}\right), \qquad (3.48)$$

which approximates the Bark-Hertz transformation specified in Schroeder (1977). In certain implementations[29] of PLP, Eq. 3.48 appears as

$$f_{Bark} = 7 \ln\left(\frac{f_{lin}}{650} + \sqrt{1 + \left(\frac{f_{lin}}{650}\right)^2}\right), \qquad (3.49)$$

which offers a more accurate approximation of the Bark scale at high frequencies.

In contrast to the various MFCC implementations that consider a filter-bank of triangular filters, the PLP speech parameterization models the critical-band masking curve with the piece-wise approximation:

$$H(f_{Bark}) = \begin{cases} 0 & \text{for} \quad f_{Bark} < -1.3 \\ 10^{2.5(f_{Bark}+0.5)} & \text{for} \quad -1.3 < f_{Bark} < -0.5 \\ 1 & \text{for} \quad -0.5 < f_{Bark} < 0.5 \\ 10^{-1(f_{Bark}-0.5)} & \text{for} \quad 0.5 < f_{Bark} < 2.5 \\ 0 & \text{for} \quad f_{Bark} > 2.5 \end{cases} \qquad (3.50)$$

---

[29] For instance, as in the VoiceBox: Speech Processing Toolbox for MATLAB, written by Mike Brookes. Available on-line at: http://www.ee.ic.ac.uk/hp/staff/dmb/voicebox/voicebox.html

Table 3.5 The PLP-FB19 filter-bank for the frequency range [0, 8000] Hz

Filter no.	Lower frequency [Hz]	Higher frequency [Hz]	Center frequency [Hz]	Filter bandwidth [Hz]
1	0	203	100	101
2	100	311	203	105
3	203	427	311	112
4	311	556	427	122
5	427	700	556	136
6	556	863	700	153
7	700	1050	863	174
8	863	1265	1050	201
9	1050	1516	1265	232
10	1265	1808	1516	271
11	1516	2150	1808	316
12	1808	2551	2150	370
13	2150	3023	2551	435
14	2551	3577	3023	511
15	3023	4231	3577	602
16	3577	5000	4231	709
17[a]	4231	5907	5000	835
18[a]	5000	6977	5907	985
19[a]	5907	8238[b]	6977	1162

[a] These filters are present only in the wideband version ([0, 8000] Hz) of the PLP filter-bank
[b] This value is truncated to 8000 Hz in the practical implementation of the filter-bank

The discrete convolution of the critical band curve with the short-time power spectrum yields samples of the critical-band warped power spectrum $Y(f_{Bark,i})$:

$$Y(f_{Bark,i}) = \sum_{f_{Bark}=-1.3}^{2.5} \left| S(f_{Bark} - f_{Bark,i}) \right|^2 H(f_{Bark}), \quad i = 1, 2, ..., M, \qquad (3.51)$$

where M is the number of filters in the equivalent filter-bank, which samples the frequency range $f_s/2$ in 1-Bark intervals. Hermansky (1990), suggested 18 spectral samples for the speech with bandwidth [0, 5000] Hz, which divides the frequency range [0, 16.9] Barks to 17 equal intervals of 0.9941 Barks each. Hermansky suggested $M = 16$ filters for covering this frequency range. In Table 3.5, we tabulate the central frequencies and the bandwidth of the filters in the filter-bank of PLP-FB19. Next, the inverse Bark-to-Hertz transformation,

$$\hat{f}_{Bark} = 600 \sinh\left(\frac{f_{Bark}}{6}\right), \qquad (3.52)$$

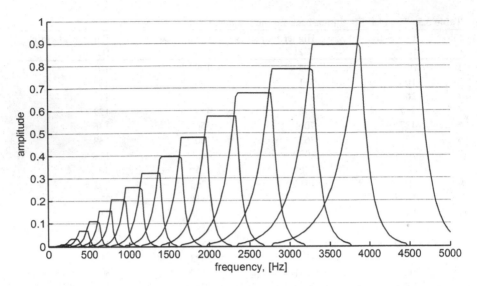

Fig. 3.8 The 16-filter PLP filter-bank for bandwidth of 5000 Hz (Hermansky 1990)

is used to convert from the Bark frequency, f_{Bark}, back to the linear frequency in Hertz, $\hat{f}_{Bark}$ that samples $Y(f_{Bark,i})$. The next step in the computation of the PLP cepstral coefficients is the equal-loudness pre-emphasis,

$$Z(\hat{f}_{Bark,i}) = E(f_{lin})Y(\hat{f}_{Bark,i}),\tag{3.53}$$

where the simulated equal-loudness curve is defined as:

$$E(f_{lin}) = \left(\frac{f_{lin}^2}{f_{lin}^2 + 1.6 \cdot 10^5}\right)^2 \cdot \left(\frac{f_{lin}^2 + 1.44 \cdot 10^6}{f_{lin}^2 + 9.61 \cdot 10^6}\right).\tag{3.54}$$

The 16-filter filter-bank defined by (3.50), scaled after applying the equal-loudness pre-emphasis (3.54), is shown in Fig. 3.8.

Afterward, a cubic-root amplitude compression is applied,

$$F(f_{Bark}) = \sqrt[3]{Z(f_{Bark})},\tag{3.55}$$

in order to simulate the nonlinear relation between the intensity of sound and its perceived loudness, and approximates the power law of hearing (Stevens 1957).

Finally, $F(f_{Bark})$ is approximated by the spectrum of an all-pole model using the autocorrelation method of all-pole spectral modeling (Hermansky 1990). Details about the all pole spectral modeling are offered in Makhoul (1975).

3.7 Comparison Among the Various DFT-Based Speech Features

In this subsection, we discuss the main differences among the various DFT-based speech parameterization methods presented in Sects. 3.2–3.6. Depending on the selected criteria, different grouping of these methods can be made.

First, with respect to the sequence of processing steps, the computation of the PLP cepstral coefficients differs significantly from the computation of the other DFT-based speech parameterizations: LFCC, MFCC, and HFCC, and the only similarity is in the use of signal windowing and the DFT. Numerous studies have reported the advantage of the PLP or the MFCC on different speech processing tasks, but it remains an application-specific problem to choose between PLP and the LFCC / MFCC / HFCC schemes. A comparison between the performance of the PLP cepstral coefficients and the other speech features considered in this book on the monophone recognition task and on the continuous speech recognition tasks is offered in Sects. 5 and 6, respectively.

Another criterion for categorizing the speech parameterizations is the filter-bank design. Figure 3.9 illustrates the filter-banks of four commonly used speech parameterizations for the frequency range [0, 8000] Hz. In particular, the HTK implementation of the MFCC with 26 filters in the frequency range [0, 8000] Hz is the default filter-bank in the recent versions of the HTK (Young et al. 2006). It is shown in Fig. 3.9 with dotted line and marker "▲." Next, the filter-bank used in the MFCC implementation of Slaney (1998), is shown with a solid line and marker "◆." The filter-bank used in the PLP (Hermansky 1990) is shown with solid line and marker "◻" and the filter-bank used in the HFCC speech parameterization (Skowronski and Harris 2004) is shown with a dashed line and marker "o."

As Fig. 3.9 presents, the filter bandwidths for these filter-banks are proportional for frequencies above 1000 Hz, except the filter-bank for HFCC, which has a different slope. As shown in the figure, there are significant differences among these filter-banks in the approximation of the critical bandwidth for the frequencies below 1000 Hz. For the lower frequency range, the HFCC uses filters with narrow bandwidth and thus offers better frequency resolution in this range when compared to the other speech parameterizations. That is not necessarily needed for the speech recognition tasks but could benefit the speaker recognition process. Furthermore, the filter-banks of HFCC and the MFCC of Slaney cover the frequency range with more filters. However, the filters used in the HFCC-FB29 are much wider for center frequencies above 500 Hz, than these used in the MFCC-FB40 speech parameterization. For small values of the E-factor, for example, $E \leq 2$, at least some of the HFCC filters with center frequency below 500 Hz remain narrower than these used in the other MFCC implementations or the PLP. In addition, in the HFCC-E scheme, the filters with the highest center frequencies overlap widely, which is in contrast to the other MFCC implementations, where the overlap is fixed by the selection of the central frequency of the left and right neighboring filters. Thus, in the HFCC speech parameterization scheme, each filter overlaps not only with its immediate neighbors, as in the other MFCC implementations, but also may overlap with several more distant neighbors.

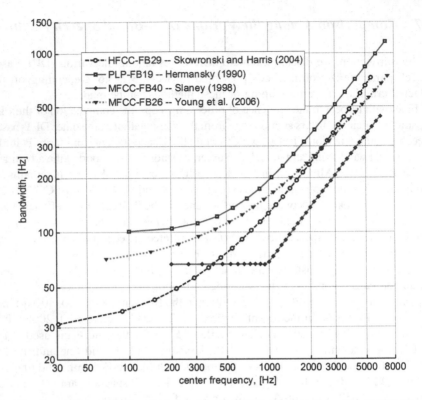

Fig. 3.9 Comparison of the filter-banks used in the HTK implementation of the MFCC (Young et al. 2006), the MFCC implementation of Slaney (1998), the PLP of Hermansky (1990), and the HFCC speech features (Skowronski and Harris 2004)

The decoupling of the filter bandwidth from the center frequency of the filters offers the opportunity for a certain adjustment of the filter-bank design for the needs of different applications. For instance, after experimenting on the isolated digits recognition task for English language, Skowronski and Harris (2004), reported that increasing the E-factor (up to $E = 5.0$) improves both the speech recognition accuracy and the noise-robustness of the HFCC-FB29 speech features. Further-more, Ganchev et al. (2005), studied the influence of the E-factor on the speaker verification accuracy, and reported that the lowest error rate is obtained for $E = 0.5$. Thus, the range of values the E-factor takes could span one order of magnitude, and shall be adjusted according to the needs of each particular applica-tion. However, in the HFCC-E scheme, the E-factor controls the bandwidth of all filters simultaneously, and the proportion between their bandwidth remains unchanged. Another interesting research direction could be to learn the filter spacing and bandwidths in data-dependent manner for each particular application.

Finally, according to our present understanding about the properties of the human auditory system, the HFCC seem the most up-to-date speech parameterization scheme among all DFT-based speech parameterizations. This is because in the HFCC the filter bandwidth is decoupled from the center frequencies of the neighboring filters, the frequency warping, and the filter bandwidth is in conformance to the insight (Moore 2003) that at low frequencies, the critical bandwidth goes well below 100 Hz, and as low as 30 Hz for center frequency 30 Hz.

With respect to their performance, the DFT-based speech parameterizations are difficult to rank in absolute terms, as their performance is highly dependant on the task, experimental setup, database, etc. In the following, we briefly summarize some indicative and conclusive studies about the performance the DFT-based speech features, according to results reported in the literature. A number of studies, starting with (Davis and Mermelstein 1980), investigated the appropriateness of various speech features for the monophone recognition and isolated word recognition tasks, and MFCC were reported to outperform the LFCC and the LPC-based speech features. Specifically, in Davis and Mermelstein (1980), it was demonstrated that MFCC outperform LPC, LPCC, and other features, on the task of isolated word recognition.

Hermansky (1990) compared the performance of the PLP speech features against the LP coefficients in two dissimilar setups: (i) single-frame monophone identification and (ii) isolated-word recognition for vocabulary of 36 words. On both setups, the PLP cepstral coefficients demonstrated an advantage over the LP parameters for the low-order model analysis (for order 2–5) but for higher orders of the model the accuracy of PLP decreased, and for the order 13 and higher the PLP lost their advantage. Later on, various authors compared the PLP cepstral coefficients and the MFCC on different speech processing tasks, and often the results were not conclusive or were contradicting to other studies on the same task.

Furthermore, on various speech processing tasks, it was demonstrated (Davis and Mermelstein 1980; Reynolds 1994; Chen et al. 1997) that in noisy conditions MFCC preserve higher robustness when compared to other speech features, such as LPCC, PLP, etc. Due to the advantageous performance and robustness that MFCC demonstrated in these early studies, the various MFCC implementations become widely used by researchers and technology developers. The MFCC were included in the ETSI standards[30,31] for speech communications and at present dominate as the common choice in nearly all speech processing tasks, which do not require reconstruction of the speech signal.

Next, experimenting with isolated digits recognition, Skowronski and Harris (2004) reported that the HFCC-E speech features offer advantageous accuracy and

---

[30] ETSI ES 201 108, V1.1.2 (2000-4). ETSI Standard: Speech Processing, Transmission and Quality Aspects (STQ); Distributed Speech Recognition; Extended Advanced Front-end Feature Extraction Algorithm; Compression Algorithms; Back-end Speech Reconstruction Algorithm, April 2000, Chapter 4, pp. 8–11.

[31] ETSI ES 202 050, V1.1.5 (2007-1). ETSI Standard: Speech Processing, Transmission and Quality Aspects (STQ); Distributed Speech Recognition; Extended Advanced Front-end Feature Extraction Algorithm; Compression Algorithms; Back-end Speech Reconstruction Algorithm; January 2007, Section 5.3, pp. 21–24.

increased noise robustness over the MFCC implementations of Davis and Mermelstein (1980), Young et al. (1995), and Slaney (1998). On the speaker verification task, Ganchev et al. (2005) reported that the HFCC-E for E-factor $E = 0.5$ outperform the other MFCC implementations in terms of equal error rate, but were outperformed by the MFCC of Slaney (1998), in terms of optimal decision cost. In addition, the HFCC-E speech features were found advantageous on the speech segmentation task, where they outperformed few other DFT-based and discrete wavelet packet transform-based speech parameterizations (Mporas et al. 2008).

Sections 5–7 offer a direct comparison of the performance of the DFT-based speech parameterizations discussed in Sect. 3 (PLP, LFCC, MFCC, and HFCC), as well as a comparison against five discrete wavelet packet transform-based speech parameterizations discussed in Sect. 4. In this comparison, we consider the monophone recognition task (Sect. 5), the speech recognition task (Sect. 6), and the speaker verification task (Sect. 7).

4 DWPT-Based Speech Parameterization

Section 4 offers a detailed description of five discrete wavelet packet transform (DWPT)-based speech parameterization methods, which were reported advantageous on various speech processing tasks. These methods are considered relatively more interesting not only because they brought new research perspectives to the area of speech parameterization, but also because they attracted some attention and were quoted in work of independent researchers. These five methods are:

- WPF-SBC – sub-band coding DWPT-based speech features (Sarikaya et al. 1998)
- WPF-FD – DWPT-based speech features (Farooq and Datta 2001)
- WPF-OBJ – objective DWPT-based speech features (Siafarikas et al. 2004, 2007)
- WPF-OVL – overlapping DWPT-based speech features (Siafarikas et al. 2005)
- WPF-ACE – DWPT-based speech features that make use of the filter-bank of Nogueira et al. (2006)

In fact, these five DWPT-based speech parameterizations share the general processing steps with the DFT-based methods discussed in Sect. 3, and take advantage of the universal speech processing concepts discussed in Sect. 1.2. In particular, the WPF-SBC, WPF-FD, WPF-OBJ, WPF-OVL, and WPF-ACE build on the idea of sub-band processing of speech, and employ frequency warping motivated by the nonlinear pitch perception in the human auditory system and the critical band concept, and also make use of the cepstral analysis of speech. However, the major difference with the DFT-based schemes, discussed in Sect. 3, is that here the time-frequency decomposition is performed through the DWPT. This offers a greater flexibility in the selection of time-frequency resolution trade-offs and allows fine-tuning of the analysis in the signal decomposition stage by careful selection of the set of basis functions. Another difference is that while the DFT-based speech parameterizations aim at achieving an accurate representation of the nonlinear perception of pitch in the human auditory system, and attempt to closely approximate the critical bandwidth of the auditory filters, the DWPT-based methods only consider these as reference concepts, as with the DWPT their accurate approximation is not feasible. The last is not necessarily a disadvantage (Sects. 6 and 7 offer evidence in support of this statement), but allows the speech parameterizations research to be given a new meaning – adaptation of the speech parameterization process to the objective needs of each target application, instead of trying to precisely approximate the human auditory system.

In the following sections, the DWPT-based speech parameterizations of interest are presented in chronological order, sorted by the year of their initial emergence in the literature.

4.1   The WPF-SBC

Erzin et al. (1995) developed a sub-band-based speech analysis scheme, which makes feasible the derivation of speech features that are less sensitive to noise. In particular, Erzin et al. implemented two speech parameterization techniques which exploit this sub-band analysis scheme: (i) sub-band-based line spectral frequency (SBLSF) parameters and (ii) sub-band-based cepstral coefficients (SUBCEP). The performance of the SBLSF and SUBCEP was evaluated on the isolated word recognition task, involving a vocabulary of ten digits in the presence of car noise and a range of SNR conditions. Both the SBLSF and SUBCEP were reported to outperform significantly the widely used LSF and MFCC speech parameters.

In brief, the SUBCEP parameters are computed by passing the speech signal through a perfect reconstruction filter-bank with 22 FIR filers, followed by a DCT-based decorrelation of the filter-bank outputs. The FIR filter-bank roughly approximates the Mel scale. Later on, Sarikaya et al. (1998), re-implemented this filter-bank by using Daubechies' 32-tap orthogonal filters and employed a slightly different approximation of the Mel scale, using 24 bands for the frequency range [0, 4000] Hz, instead of the 22 used in Erzin et al. (1995). Sarikaya et al. named their set of speech features sub-band cepstral (SBC) parameters; however, in order to emphasize the use of the wavelet packet transform with the Daubechies wavelet function of order 32, we refer to these speech features as to WPF-SBC, where WPF stands for discrete wavelet packet transform-based speech features. The relation between central frequency and filter bandwidth in the filter-banks employed in the SUBCEP and the WPF-SBC is shown in Fig. 4.1, together with these used in the PLP-FB19 (Hermansky 1990) and the MFCC-FB40 (Slaney 1998).

As Fig. 4.1 shows, the filter-banks used in the computation of the SUBCEP and WPF-SBC offer slightly dissimilar approximations of the Mel scale in the frequency range [1000, 3000] Hz, and are identical for frequencies below 1000 Hz and over 3000 Hz. In the range [1000, 3000] Hz, the WPF-SBC filter-bank offers slightly better frequency resolution when compared to the SUBCEP filter-bank. Both the filter-banks emphasize low-to-mid frequencies assigning more sub-bands

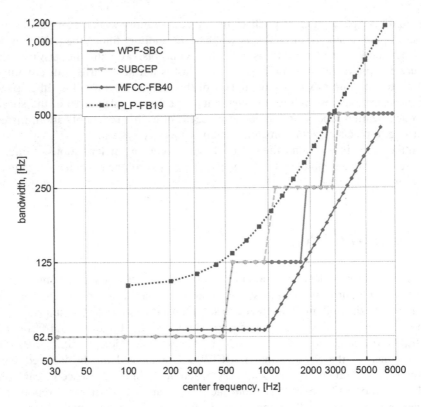

Fig. 4.1 The relation between the center frequency and bandwidth of the filters employed in the filter-banks of the WPF-SBC, SUBCEP, MFCC-FB40, and the PLP-FB19 speech features

in this range, and as a whole, their decomposition preserves roughly logarithmic distribution of the sub-bands across frequency, which resembles the properties of the human auditory apparatus. In particular, for frequencies below 500 Hz, the filter-banks of the SUBCEP and WPF-SBC offer frequency resolution comparable to the one of Slaney's approximation of the Mel scale, and in the entire frequency range [0, 4000] Hz they offer a finer (higher) resolution than the Bark-spaced filter-bank used in the PLP. To this end, a direct comparison of the performance of the SUBCEP and WPF-SBC filter-banks has not been published, and thus it remains unknown which one is more appropriate in the various speech processing tasks. However, in Sarikaya et al. (1998) there is the implicit proposition that the filter-bank with 24 filters for the frequency range [0, 4000] Hz is more appropriate for the needs of speaker identification than some other candidates they evaluated.

Although in this book a direct comparison between these two filter-banks is not pursued, in the experimental evaluation of various implementations of the Mel scale on the speaker verification task (Sect. 7), we comment on the empirical evidence that a filter-bank with more filters aids for higher speaker recognition accuracy. This indirectly supports the statement of Sarikaya et al. (1998) that their filter-bank is advantageous over the one in Erzin et al. (1995).

The computation of the WPF-SBC parameters is summarized as follows. Let us denote with n the discrete-time index, and with $x(n)$ a discrete-time speech signal that has been sampled with sampling frequency f_s and therefore has spectral content bounded in the frequency range $[0, 0.5] f_s$. Let us consider that the signal $x(n)$ has been pre-processed as explained in Sect. 2.2, which results in the pre-emphasized signal $s(n)$, weighted by a rectangular window with length of N samples, where $N = 2^J$ is an exact power of two for some positive integer J. By applying the DWPT, the speech signal, $s(n)$, is partitioned (Siafarikas et al. 2007) as:

$$\mathbf{W}_j^{2n}(k) = \sum_{i=0}^{L-1} a_{n,i} \mathbf{W}_{j-1}^n \left[(2k+1-i) \bmod \left(N/2^{j-1} \right) \right], \qquad (4.1)$$

$$\mathbf{W}_j^{2n+1}(k) = \sum_{i=0}^{L-1} b_{n,i} \mathbf{W}_{j-1}^n \left[(2k+1-i) \bmod \left(N/2^{j-1} \right) \right], \qquad (4.2)$$

where $k = 0, 1, \ldots, N/2^j - 1$, and L is the order of the wavelet function. The parent signal on the top of the decomposition tree is $\mathbf{W}_0^0(n) \equiv s(n)$, and the wavelet filter h_i and the scaling filter g_i represented by the coefficients $a_{n,i}$ and $b_{n,i}$ are applied depending on the position in the decomposition tree as follows:

(i)   For the *even* values of n, $a_{n,i} \equiv g_i$ and $b_{n,i} \equiv h_i$
(ii) For the *odd* values of n, $a_{n,i} \equiv h_i$ and $b_{n,i} \equiv g_i$

By going from level $j - 1$ to the next decomposition level j, each *parent node* $\mathbf{W}_{j-1}^{n'}$ of the decomposition tree is circularly filtered and down-sampled twice: once with the wavelet filter $\{h_l\}$ and once with the scaling filter $\{g_l\}$, yielding two *children nodes* $\mathbf{W}_j^n$, indexed by $n = 2n'$ and $n = 2n' + 1$. Each partition of the original signal, $\mathbf{W}_j^n$, obtained in this manner contains the information for the frequency interval $[n/2^{j+1}, (n+1)/2^{j+1}]$ and provides information associated with the time interval $[2^j k, 2^j (k+1)]$. Once the children nodes of interest are computed, the corresponding coefficient vectors, $\mathbf{W}_j^n$, are stacked together to form an orthogonal subset $S = \{ \mathbf{W}_j^n : j = 0, 1, \ldots, J, n = 0, 1, \ldots, 2^j - 1 \}$, which corresponds to the frequency division defined by the desired decomposition tree. The WPF-SBC makes use of DWPT with Daubechies' wavelet filter of order 32.

Initially, the WPF-SBC were defined (Sarikaya et al. 1998) for sampling frequency $f_s = 8000$ Hz, and frame size of 24 ms, which corresponds to $N = 194$ speech samples. However, for the purpose of fair comparison with the other speech parameterization schemes considered in this book, we will consider frame size of 32 ms, which for sampling rate of $f_s = 8000$ Hz, results to $N = 256$ samples. The frequency division with 24 sub-bands, covering the frequency range $[0, 4000]$ Hz, is summarized in Table 4.1. This frequency resolution is implemented with a wavelet packet tree of depth $D = 6$, defined with the orthogonal subset $S_{SBC,8kHz} = \{ [\mathbf{W}_6^0, \mathbf{W}_6^7], [\mathbf{W}_5^4, \mathbf{W}_5^{13}], [\mathbf{W}_4^7, \mathbf{W}_4^9], [\mathbf{W}_3^5, \mathbf{W}_3^7] \}$. The wavelet packet decomposition, which the subset $S_{SBC,8kHz}$ represents, is illustrated in Fig. 4.2.

Table 4.1 Bandwidth and number of filters in the original and wideband version of the WPF-SBC filter-bank

Frequency range [Hz]	Number of sub-bands	Bandwidth [Hz]
[0, 500]	8	62.5
[500, 1750]	10	125
[1750, 2500]	3	250
[2500, 4000]	3	500
[4000, 8000][a]	8	500

[a] The eight sub-bands covering the frequency range [4000, 8000] Hz exist only for the wideband version of the WPF-SBC speech features

Furthermore, let us also consider a wideband version of the WPF-SBC speech features, which are computed for signals sampled at frequency $f_s = 16000$ Hz, that is, speech bandwidth of [0, 8000] Hz. As in this case the bandwidth of the original signal is twice wider, and in order to preserve the frequency resolution defined in the original filter-bank one needs to increase the depth of the wavelet packet decomposition by one. Thus, the DWPT shall be performed up to depth $D = 7$. Next, in order to adapt the original WPF-SBC filter-bank for the frequency range [0, 8000] Hz, eight new sub-bands, with bandwidth of 500 Hz each are added to cover the frequency range [4000, 8000] Hz. The last preserves the Mel-scale-like frequency warping and leads to the actual frequency range of [0, 8000] Hz that is covered by 32 frequency sub-bands as specified in Table 4.1. In this manner, the corresponding orthogonal subset of wavelet coefficients used in the wideband version of the WPF-SBC parameters is defined as $S_{SBC,16kHz} = \{[\mathbf{W}_7^0, \mathbf{W}_7^7], [\mathbf{W}_6^4, \mathbf{W}_6^{13}], [\mathbf{W}_5^7, \mathbf{W}_5^9], [\mathbf{W}_4^5, \mathbf{W}_4^{15}]\}$.

The wavelet packet decomposition obtained to this end resulted in the selection of p disjoint sub-bands, defined as $S_{SBC} \in \{S_{SBC,8kHz}, S_{SBC,16kHz}\}$. Next, the energy E_p in each of the B sub-bands, scaled by the number of coefficients, $N_p = N/2^j$, in the corresponding sub-band, is computed as:

$$E_p = \frac{1}{N_p} \sum_{m=1}^{N_p} \left[\mathbf{W}_j^n(m)\right]^2, \quad \mathbf{W}_j^n \in S_{SBC}, \quad p = 1, 2, ..., B, \quad (4.3)$$

where $\mathbf{W}_j^n(m)$ is the mth individual coefficient of the DWPT vector at the specific node $\mathbf{W}_j^n \in S_{SBC}$, and j is the depth of wavelet decomposition at which the specific coefficient vector is located. The sub-band signal energies, E_p, obtained in this manner are next compressed logarithmically and then transformed to cepstral coefficients through the DCT applied on the log-energies, as:

$$WPFSBC(r) = \sum_{p=1}^{B} \log_{10}(E_p) \cos\left(\frac{r(p - 0.5)\pi}{B}\right), \quad r = 0, 1, ..., R - 1, \quad (4.4)$$

where B is the total number of frequency sub-bands and $R \leq B$ is the number of unique WPF-SBC cepstral coefficients that can be computed. For larger R, the

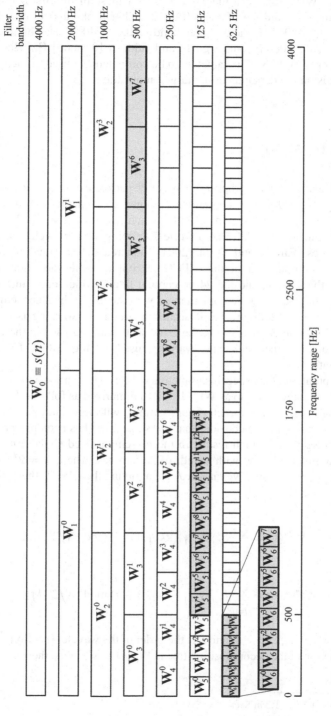

Fig. 4.2 Wavelet packet decomposition $S_{SBC,8kHz}$ for the original bandwidth ([0, 4000] Hz) of the WPF-SBC speech features

values of the WPF-SBC with index $r \geq B$ mirror these of the first B coefficients. According to the definition of the WPF-SBC filter-banks above, one may have $R \leq 24$, for $f_s = 8000$ Hz, and signal bandwidth $[0, 4000]$ Hz, or alternatively $R \leq 32$, for $f_s = 16000$ Hz and signal bandwidth of $[0, 8000]$ Hz. However, the actual number of WPF-SBC parameters to be computed and used in each particular setup depends on the demands of the target application.

4.2 The WPF-FD

Farooq and Datta (2001), proposed a wavelet packet decomposition of the frequency range $[0, 8000]$ Hz such that the obtained 24 frequency sub-bands closely follow the Koenig scale[32] (Koenig 1949), which is exactly linear below 1000 Hz and logarithmic above 1000 Hz. In brief, defining their wavelet packet-based speech features, Farooq and Datta (2001), adhered to the standard processing steps for the computation of the MFCC, except that the short-time DFT was replaced by the DWPT, and the 24 perceptual filters, which constitute a wavelet packet-based filter-bank, were implemented with the use of the Daubechies' wavelet function of order 12. Figure 4.3 illustrates the frequency warping used by Farooq and Datta (2001), in comparison to the Bark-scale warping used in the PLP-FB19 filter-bank and to the Mel-scale warping used in the MFCC-FB40 speech parameterization.

The computation of the speech parameters proposed by Farooq and Datta (2001), which in this book we refer to as WPF-FD, is summarized as follows. Let us denote with $x(n)$ the discrete-time speech signal, sampled with frequency f_s, where n is the discrete time index. Let us consider that the signal $x(n)$ has been pre-processed as explained in Sect. 2.2, which results in the pre-emphasized signal $s(n)$, weighted with a rectangular window with length of N samples, where $N = 2^J$ is an exact power of two for some positive integer J. By applying the DWPT, the speech signal $s(n)$ is partitioned as:

$$\mathbf{W}_j^{2n}(k) = \sum_{i=0}^{L-1} a_{n,i} \mathbf{W}_{j-1}^n \left[(2k + 1 - i) \bmod \left(N/2^{j-1} \right) \right], \tag{4.5}$$

$$\mathbf{W}_j^{2n+1}(k) = \sum_{i=0}^{L-1} b_{n,i} \mathbf{W}_{j-1}^n \left[(2k + 1 - i) \bmod \left(N/2^{j-1} \right) \right], \tag{4.6}$$

where $k = 0, 1, \ldots, N/2^j - 1$, and L is the order of the wavelet function. The parent signal on the top of the decomposition tree is $\mathbf{W}_0^0(n) \equiv s(n)$, and the wavelet filter h_i

---

[32] Refer to Eqs. 1.1 and 1.2 in Sect. 1.3.

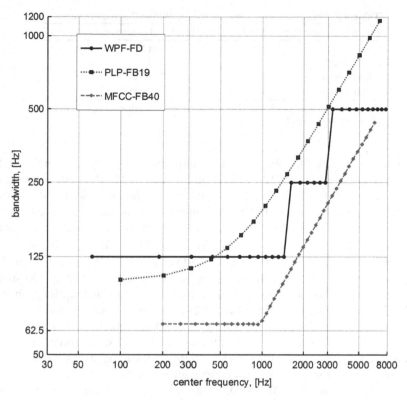

Fig. 4.3 The frequency warping used in the WPF-FD (*solid line with mark* "●") in comparison to the Bark-scale warping used in PLP-FB19 (*dotted line with mark* "■") and the Mel-scale warping used in the MFCC-FB40 (*dashed line with mark* "◆") speech features

and the scaling filter g_i represented by the coefficients $a_{n,i}$ and $b_{n,i}$ are applied depending on their position in the decomposition tree, as follows:

(i) For the *even* values of n, $a_{n,i} \equiv g_i$ and $b_{n,i} \equiv h_i$
(ii) For the *odd* values of n, $a_{n,i} \equiv h_i$ and $b_{n,i} \equiv g_i$

By going from level $j - 1$ to the next decomposition level j, each *parent node* $\mathbf{W}_{j-1}^{n'}$ of the decomposition tree is circularly filtered and down-sampled twice: once with the wavelet filter $\{h_l\}$ and once with the scaling filter $\{g_l\}$, yielding two *children nodes* $\mathbf{W}_j^n$, indexed by $n = 2n'$ and $n = 2n' + 1$. Each partition, $\mathbf{W}_j^n$, of the original signal $\mathbf{W}_0^0(n)$, obtained in this way contains the information for the frequency interval $[n/2^{j+1}, (n+1)/2^{j+1}]$ and provides information associated with the time interval $[2^j k, 2^j(k+1)]$. Once all the children nodes of interest are computed, the corresponding coefficient vectors $\mathbf{W}_j^n$ are stacked together to form an orthogonal subset $S = \{\mathbf{W}_j^n : j = 0, 1, \ldots, J, n = 0, 1, \ldots, 2^j - 1\}$, which corresponds to the frequency division defined by the desired tree. In the computation of the WPF-FD, the DWPT is implemented by means of Daubechies' wavelet filter of order 12.

Table 4.2 Frequency decomposition for the WPF-FD speech features

Frequency range [Hz]	Number of sub-bands	Frequency resolution [Hz]
[0, 1500]	12	125
[1500, 3000]	6	250
[3000, 4000]	2	500
[4000, 8000][a]	4	1000

[a] These sub-bands exist only for the wideband version, signal bandwidth [0, 8000] Hz of the WPF-FD features

Initially, the WPF-FD were defined for sampling frequency $f_s = 16000$ Hz and frame size of 32 ms ($N = 512$), which corresponds to 24 sub-bands covering the frequency range [0, 8000] Hz. For signal bandwidth of 8000 Hz the desired frequency resolution (Table 4.2) is implemented by means of a wavelet packet decomposition tree of depth $D = 6$, which leads to the orthogonal subset $S_{FD,16kHz} = \{[\mathbf{W}_6^0, \mathbf{W}_6^{11}], [\mathbf{W}_5^6, \mathbf{W}_5^{11}], [\mathbf{W}_4^6, \mathbf{W}_4^7], [\mathbf{W}_3^4, \mathbf{W}_3^7]\}$, shown on Fig. 4.4.

In addition, let us also define a narrowband version of the WPF-FD for sampling rate $f_s = 8000$ Hz, where the frame size of 32 ms corresponds to $N = 256$ speech samples. Since in this case the signal content is restricted to the frequency range [0, 4000] Hz, the last four sub-bands (Table 4.2) are excluded from the wavelet packet decomposition tree. Furthermore, as for $f_s = 8000$ the signal $s(n)$ has only half of the bandwidth and only half of the samples for equal frame size, in order to preserve the original resolution of the filter-bank, as defined in Farooq and Datta (2001), one needs to decrease the depth of the wavelet packet tree to $D = 5$. Therefore, for the narrowband version of the WPF-FD parameters, we make use of the orthogonal subset $S_{FD,8kHz} = \{[\mathbf{W}_5^0, \mathbf{W}_5^{11}], [\mathbf{W}_4^6, \mathbf{W}_4^{11}], [\mathbf{W}_3^6, \mathbf{W}_3^7]\}$, which covers the frequency range [0, 4000] Hz.

Following the wavelet packet decomposition, $S_{FD} \in \{S_{FD,8kHz}, S_{FD,16kHz}\}$, in total B disjoint sub-bands are formed. The energy E_p in the pth sub-band is calculated as:

$$E_p = \frac{1}{N_p} \sum_{m=1}^{N_p} \left[\mathbf{W}_j^n(m)\right]^2, \quad \mathbf{W}_j^n \in S_{FD}, \quad p = 1, 2, ..., B, \qquad (4.7)$$

where $\mathbf{W}_j^n(m)$ is the mth individual coefficient of the DWPT vector at the specific node $\mathbf{W}_j^n$, j is the decomposition level at which the coefficient vector is located, p is the sub-band index. Here, the energy E_p in each sub-band is normalized with the number of wavelet packet coefficients N_p in the corresponding sub-band.

Subsequently, the normalized sub-band energies (4.7) obtained at the output of the filter-bank are compressed logarithmically and decorrelated by the DCT, as:

$$WPFFD(r) = \sum_{p=1}^{B} \log_{10}(E_p) \cos\left(\frac{r(p - 0.5)\pi}{B}\right), \quad r = 0, 1, ..., R - 1. \qquad (4.8)$$

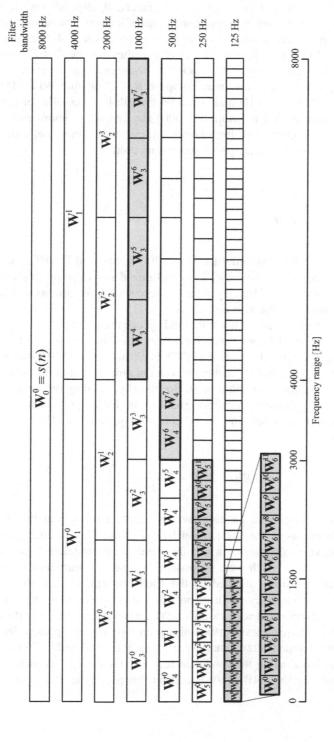

Fig. 4.4 Wavelet packet decomposition for the original bandwidth of the WPF-FD speech parameterization ([0, 8000] Hz)

Here, the total number of frequency sub-bands, B, depends on the selected sampling frequency, f_s, and the signal bandwidth of interest. The number of unique WPF-FD cepstral coefficients which can be computed is $R \leq B$. For larger R, the values of the WPF-FD with index $r \geq B$ mirror these of the first B coefficients. According to the wavelet packet decomposition defined above, $S_{FD} \in \{S_{FD,8kHz}, S_{FD,16kHz}\}$, one may compute $R \leq 24$ unique WPF-FD cepstral coefficients for $f_s = 16000$ Hz and signal bandwidth [0, 8000] Hz, or $R \leq 20$ for $f_s = 8000$ Hz and signal bandwidth [0, 4000] Hz. The actual number of WPF-FD parameters to be computed for certain experimental setup depends on the requirements of the particular speech processing task.

4.3   The WPF-OBJ

Siafarikas et al. (2004) and Siafarikas et al. (2007) presented DWPT-based speech parameterizations, which are objectively optimized (to a certain degree) for the speaker verification task. Here, we refer to these speech features as to WPF-OBJ, where OBJ stands for objective optimization.

In brief, the computation of the WPF-OBJ speech features follows the processing steps typical for the DFT-based speech parameterizations discussed in Sect. 3. The main differences are that the short-time DFT is replaced by a DWPT-based decomposition with use of the Battle-Lemarié wavelet of order 5, and in the quite different motivation and design of the filter-bank. Specifically, building on the critical band concept and on the equivalent rectangular bandwidth (ERB) approximation of critical bandwidth, shown as a function of the center frequency f in Hz (Glasberg and Moore 1990),

$$ERB = 24.7 \left(4.37 \frac{f}{10^3} + 1 \right), \tag{4.9}$$

Siafarikas et al. employed an objective mechanism for tuning the filter-bank design for the needs of the speaker verification task. This mechanism is based on the objective evaluation of a certain number of candidate wavelet packet decomposition trees, which were purposely designed to bear some resemblance to the frequency warping defined by the ERB concept. In fact, the ERB as a function of center frequency was used only as a reference around which Siafarikas et al. (2007) experimented with various DWPT resolutions. The actual frequency resolution in all frequency sub-bands of a prototype wavelet packet decomposition tree were decided after extensive experimentations with different DWPT resolutions and evaluating the candidate resolution a step higher or lower below these suggested by the ERB (4.9). Based on this prototype decomposition tree,

16 candidate wavelet packet trees were designed for the frequency range [0, 4000] Hz. These candidate trees differ in the approximation of the ERB in the frequency sub-bands [1000, 1250] Hz and [2250, 2750] Hz, and are identical in the remaining sub-bands. A systematic objective assessment of the appropriateness of these 16 candidate wavelet packet decomposition trees for the speaker verification task was performed in a common experimental setup.[33] The measure for appropriateness of a certain decomposition tree was the classification error calculated at the equal error rate (EER) point, that is, where the errors of wrongly rejecting trials belonging to the true target speaker (missing the target) and wrongly accepting nontarget trials as originating from the true speaker (false alarm) are equal.[34] Thus, the specific bandwidths, which are used in the WPF-OBJ filter-bank, resulted as an outcome of the objective evaluation of multiple candidate resolutions with respect to the speaker verification performance obtained by an automatic system.[35]

The experimentations described above resulted to a filter-bank that consists of 68 sub-bands, which cover the frequency range [0, 4000] Hz. Thus, the number of sub-bands in the WPF-OBJ filter-bank is approximately three times the number of filters in the filter-banks used in the MFCC-FB20 and the HTK MFCC-FB24.

Figure 4.5 illustrates the WPF-OBJ filter-bank in comparison to the ERB values (Moore 2003) for the same frequencies, as the center points of each sub-band from the WPF-OBJ filter-bank. As Fig. 4.5 shows, the WPF-OBJ filter-bank has significantly smaller bandwidth of the filters, when compared to the values suggested by the ERB (Eq. 4.9), especially for filters with high center frequencies. Based on this result, and on the comparison between the ERB values and the most advantageous bandwidth division obtained for the WPF-OBJ filter-bank, one can infer that human

---

[33] This experimental setup was based on the use of a particular speaker verification system with known performance (Ganchev et al. 2002a, b) and on the Polycost database (Hennebert et al. 2000). Afterward, the wavelet packet decomposition obtained in this setup was validated on the quite dissimilar NIST 2001 speaker recognition database (NIST 2001), where the advantage of the WPF-OBJ speech features over other DFT-based and wavelet packet-based speech parameterizations was confirmed (Siafarikas et al. 2007).

[34] Section 7.1 offers details on these and other performance measures used on the speaker verification task.

[35] Here, we have to point out that Siafarikas et al. (2007) did not perform an exhaustive search of all feasible filter bandwidths for the WPF-OBJ filter-bank but by practical reasons limited their search to these DWPT resolutions which are close to the values suggested by the ERB approximation of Moore (2003). Thus, the existence of a filter-bank that is even better tuned for the speaker recognition tasks, than the filter-bank currently used in WPF-OBJ, is quite probable.

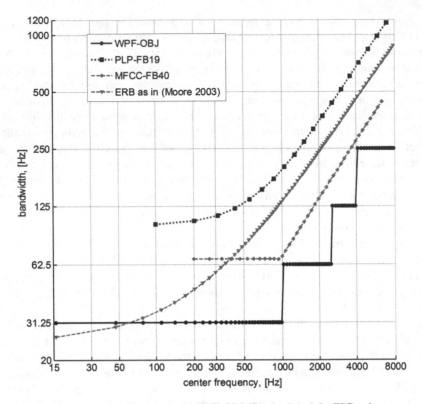

Fig. 4.5 The frequency warping used in the WPF-OBJ filter-bank and the ERB values computed (Moore 2003) for the same center frequencies in comparison to frequency warping used in the MFCC-FB40 and PLP-FB19 filter-banks

auditory apparatus (and its ERB approximation) might not provide the optimal frequency resolution for achieving outstanding speaker recognition accuracy[36] (if the speaker recognition performance is based only on short-time spectral clues).

---

[36] Recent evaluation results (Greenberg et al. 2010) indicate that for a large number of unfamiliar speakers, technology outperforms humans on the speaker verification task. Although the advantage of technology shall not be attributed solely to the limited frequency resolution of the human auditory apparatus, it might also contribute to these results. Furthermore, one possible explanation for the relatively restricted capability of humans to distinguish between similar voices, when compared to technology, could be linked to the insights offered in LePage (2003). In brief, LePage pointed out that the map curvature in mammalian cochlea depends on the trade-off between optimal frequency resolution and auditory range. For the auditory range of human cochlea (~20000 Hz) the trade-off between the four conflicting constraints: "(i) enhancing high-frequency resolution; (ii) setting lower bound on loss of low-frequency resolution; (iii) minimizing map nonuniformity; and (iv) keeping the whole map smooth," does not favor very narrow critical bands, which therefore contributes to the limited human capacity to perceive the very fine details of speech. However, humans compensate for these limitations of the auditory system by exploiting the higher levels of information present in speech (such as prosodic, linguistic, extralinguistic clues), and for a small set of close family members or long-time friends, humans still outperform technology on the speaker recognition tasks.

The WPF-OBJ speech features are computed as follows. Let us denote with $x(n)$ the discrete-time speech signal, sampled with frequency f_s, where n is the discrete time index. Let us consider that the signal $x(n)$ has been pre-processed as explained in Sect. 2.2, which resulted in the pre-emphasized signal $s(n)$, weighted with a rectangular windows with length of N samples, where $N = 2^J$ is an exact power of two for some positive integer J. By applying the DWPT, the speech signal, $s(n)$, is partitioned as:

$$\mathbf{W}_j^{2n}(k) = \sum_{i=0}^{L-1} a_{n,i} \mathbf{W}_{j-1}^n \left[(2k + 1 - i) \bmod (N/2^{j-1}) \right], \tag{4.10}$$

$$\mathbf{W}_j^{2n+1}(k) = \sum_{i=0}^{L-1} b_{n,i} \mathbf{W}_{j-1}^n \left[(2k + 1 - i) \bmod (N/2^{j-1}) \right], \tag{4.11}$$

where $k = 0, 1, \ldots, N/2^j - 1$, and L is the order of the wavelet function. The parent signal on the top of the decomposition tree is $\mathbf{W}_0^0(n) \equiv s(n)$, and the wavelet filter h_i and the scaling filter g_i represented by the coefficients $a_{n,i}$ and $b_{n,i}$ are applied depending on their position in the decomposition tree, as follows:

(i) For the *even* values of n, $a_{n,i} \equiv g_i$ and $b_{n,i} \equiv h_i$
(ii) For the *odd* values of n, $a_{n,i} \equiv h_i$ and $b_{n,i} \equiv g_i$

By going from level $j - 1$ to the next decomposition level j, each *parent node* $\mathbf{W}_{j-1}^{n'}$ of the decomposition tree is circularly filtered and down-sampled twice: once with the wavelet filter $\{h_l\}$ and once with the scaling filter $\{g_l\}$, yielding two *children nodes* $\mathbf{W}_j^n$, indexed by $n = 2n'$ and $n = 2n' + 1$. Each one of the 2^j partitions, $\mathbf{W}_j^n$, of the original signal $\mathbf{W}_0^0(n)$, obtained in this way contains the information for the frequency interval $[n/2^{j+1}, (n+1)/2^{j+1}]$, and provides information associated with the time interval $[2^j k, 2^j (k+1)]$. Once all the children nodes of interest are computed, the corresponding coefficient vectors $\mathbf{W}_j^n$ are stacked together to form an orthogonal subset $S = \{\mathbf{W}_j^n : j = 0, 1, \ldots, J, \quad n = 0, 1, \ldots, 2^j - 1\}$, which corresponds to the frequency division defined by the desired decomposition tree. In the computation of the WPF-OBJ, the DWPT is implemented by means of the Battle-Lemarié wavelet of order 5, which was found advantageous over a number of other wavelet functions that were evaluated.

The frequency division used in the WPF-OBJ speech features (Siafarikas et al. 2007), with $B = 68$ sub-bands, covering the frequency range [0, 4000] Hz, is summarized in Table 4.3. For signal bandwidth of 4000 Hz the desired frequency resolution is implemented with the use of a wavelet packet decomposition tree with depth $D = 7$, which leads to the orthogonal subset $S_{OBJ,8kHz} = \{[\mathbf{W}_7^0, \mathbf{W}_7^{31}], [\mathbf{W}_6^{16}, \mathbf{W}_6^{39}], [\mathbf{W}_5^{20}, \mathbf{W}_5^{31}]\}$, shown in Fig. 4.6. In practical speech applications, the lowest four wavelet coefficient vectors, $[\mathbf{W}_7^0, \mathbf{W}_7^3]$, are discarded because the sub-bands in the frequency range [0, 125] Hz usually contribute little speech-related information, and thus the number of sub-bands becomes $B = 64$.

Table 4.3 Frequency resolution for the WPF-OBJ filter-banks

Frequency range [Hz]	Number of sub-bands	Frequency resolution [Hz]
[0, 1000]	32	31.25
[1000, 2500]	24	62.50
[2500, 4000]	12	125
[4000, 8000][a,b]	32	125
[4000, 8000][a,c]	16	250

[a] These sub-bands exist only for the wideband version [0, 8000] Hz of the WPF-OBJ features. Here, we consider two alternative extensions: (i) [b] with 32 sub-bands with bandwidth of 125 Hz, and (ii) [c] with 16 sub-bands each with bandwidth of 250 Hz

Furthermore, for the purpose of fair comparison with other speech features on the various speech processing tasks let us also consider a wideband version of the WPF-OBJ filter-bank, extended to the frequency range [0, 8000] Hz.

In order to adapt for the signal bandwidth of 8000 Hz, first, the depth of the DWPT decomposition is increased to $D = 8$, in order to preserve the resolution of the original wavelet packet decomposition tree unchanged. Second, the wideband version of the filter-bank is obtained from the original by adding extra sub-bands that cover the frequency range [4000, 8000] Hz. This could be implemented by complementing the original tree with 32 additional sub-bands, each with bandwidth of 125 Hz. This corresponds to a wavelet packet tree defined by the orthogonal subset $S_{OBJ125,16kHz} = \{[\mathbf{W}_8^0, \mathbf{W}_8^{31}], [\mathbf{W}_7^{16}, \mathbf{W}_7^{39}], [\mathbf{W}_6^{20}, \mathbf{W}_6^{63}]\}$, which consists of $B = 100$ frequency sub-bands. The speech features derived with this decomposition are referred to as WPF-OBJ 125. An alternative implementation is to cover the frequency range [4000, 8000] Hz with 16 sub-bands, each with bandwidth of 250 Hz. In this second case, the wavelet packet decomposition tree results in a total of $B = 84$ sub-bands, $S_{OBJ250,16kHz} = \{[\mathbf{W}_8^0, \mathbf{W}_8^{31}], [\mathbf{W}_7^{16}, \mathbf{W}_7^{39}], [\mathbf{W}_6^{20}, \mathbf{W}_6^{31}], [\mathbf{W}_5^{16}, \mathbf{W}_5^{31}]\}$, and the resulting speech features are referred to as WPF-OBJ 250.

Following the wavelet packet decomposition, defined by $S_{OBJ} \in \{S_{OBJ,8kHz}, S_{OBJ125,16kHz}, S_{OBJ250,16kHz}\}$, in total B disjoint sub-bands are formed. The energy E_p in each sub-band p is calculated as:

$$E_p = \frac{1}{N_P} \sum_{m=1}^{N_p} \left[\mathbf{W}_j^n(m)\right]^2, \mathbf{W}_j^n \in S_{OBJ}, p = 1, 2, ..., B, \quad (4.12)$$

where $\mathbf{W}_j^n(m)$ is the mth individual coefficient of the DWPT vector at the specific node $\mathbf{W}_j^n$, j is the depth level at which the coefficient vector is located, and p is the sub-band index. The energy E_p in each sub-band is normalized with the number of wavelet packet coefficients, $N_p = N/2^j$, in the corresponding sub-band. Subsequently, the normalized sub-band energies obtained at the filter-bank output are logarithmically compressed and decorrelated by applying the DCT:

$$WPFOBJ(r) = \sqrt{\frac{2}{B}} \sum_{p=0}^{B-1} \log_{10}(E_{p+1}) \cos\left(\frac{r(p+0.5)\pi}{B}\right), \quad r = 0, 1, ..., R-1, \quad (4.13)$$

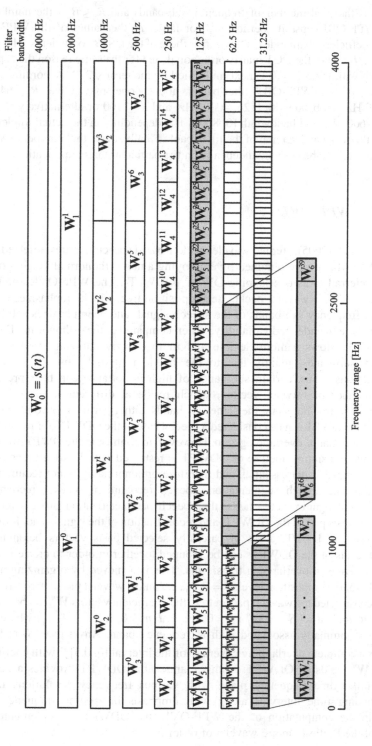

Fig. 4.6 Wavelet packet decomposition used in the WPF-OBJ speech features for signal bandwidth [0, 4000] Hz

where B is the total number of frequency sub-bands and $R \leq B$ is the number of unique WPF-OBJ cepstral coefficients. For larger R, the values of the WPF-OBJ cepstral coefficients with index $r \geq B$ mirror these of the first B coefficients. The scale factor $\sqrt{2/B}$ makes the DCT matrix orthogonal, and for the same reason the cepstral coefficient with index $r = 0$ is next multiplied by the term $\sqrt{2}/2$. According to the definition of the WPF-OBJ filter-banks above, one may have $R \leq 64$ for $f_s = 8000$ Hz and bandwidth [125, 4000] Hz, and $R \leq 100$ or alternatively $R \leq 84$ for $f_s = 16000$ Hz and bandwidth [0, 8000] Hz, depending on the actual implementation of the wideband version of the filter-bank. Finally, the actual number of WPF-OBJ parameters to be computed depends on the particular target application.

4.4   The WPF-OVL

Siafarikas et al. (2005), studied wavelet packet-based speech features, referred to as WPF-OVL, which were obtained by employing a non-orthonormal version of the DWPT, referred to as overlapping DWPT (ODWPT). The WPF-OVL employ a purposely designed wavelet packet decomposition tree, which emphasizes certain important frequency sub-bands of the speech signal, and deemphasizes sub-bands that contribute to undesired variability. When compared to the orthonormal DWPT, the ODWPT allows a more flexible and more effective utilization of specific frequency sub-bands, mainly due to the freedom to make use of redundant representations. Thus, certain sub-bands of high importance can be represented with a number of wavelet vectors, which reside at different levels of signal decomposition tree but share the same sub-bands, either partially or completely.

In fact, the DWPT can be considered a special case of the ODWPT for redundancy factor $M = 0$, that is, overlapping zero. However, in contrast to the DWPT and other orthonormal transforms, the use of ODWPT is restricted to tasks where reconstruction of the signal is not essential, and therefore orthonormality is not required. For instance, in the speech segmentation, speech recognition, speaker recognition, monophone recognition, etc., tasks, the speech parameterization process requires only the analysis part of the ODWPT, and reconstruction of the signal is not foreseen.

Likewise the DWPT, a set of carefully selected basis vectors belonging to different levels of the ODWPTs can be grouped together in order to create an even larger collection of overlapping transforms. This is achieved by organizing all the ODWPTs for decomposition levels $j = 0, 1, \ldots J$ into a wavelet packet tree structure. Having constructed the wavelet packet tree, the coefficient vectors $\mathbf{W}_j^n$ can be stacked together to form a set $S = \{\mathbf{W}_j^n : j = 0, 1, \ldots, J, n = 0, 1, \ldots, 2^j - 1\}$, where each $\mathbf{W}_j^n \in S$ is nominally associated with a frequency band. Any subset $S_1 \subset S$ that provides a complete overlapping coverage of the interval $[0, 0.5] f_s$ with coefficient vectors $\mathbf{W}_j^n$ yields an ODWPT. In this manner, the ODWPT provides a flexible tiling of the time–frequency plane with various frequency resolutions in the corresponding time intervals along with emphasis in specific frequency sub-bands. In the computation of the WPF-OVL, the ODWPT is implemented by means of the Battle-Lemarié wavelet of order 5.

Table 4.4 Frequency resolution for the WPF-OVL filter-bank

Frequency range [Hz]	Number of sub-bands	Frequency resolution [Hz]
[0, 1000]	32	31.25
[875, 1500]	10	62.50
[1500, 2000]	4	125
[2000, 2625]	10	62.50
[2375, 3000]	5	125
[3000, 3500]	8	62.50
[3500, 4000]	4	125
[4000, 8000][a]	32	125

[a] These sub-bands exist only for the wideband version [0, 8000] Hz of the WPF-OVL wavelet packet decomposition

In the design of the WPF-OVL filter-bank, Siafarikas et al. (2005) made use of a mechanism, which is very similar to the one used in the design of the sister filter-bank denoted as WPF-OBJ (Siafarikas et al. 2007), outlined in the previous Sect. 4.3. In brief, in a purposely designed test-bed on the Polycost database (Hennebert et al. 2000), Siafarikas et al. (2005) evaluated numerous candidate filter-banks, and evaluated their appropriateness for the speaker verification task. The evaluation criterion for comparison among these filter-banks was the equal error rate[37] obtained for the resultant wavelet packet speech features.

In order to restrict the number of potential candidates, Siafarikas et al. (2005) set two criteria for the selection of the candidate wavelet packet trees: First, the search was focused in these areas of the frequency axis that are considered relatively more interesting, and thus, they deserve a special care in the feature extraction process. The sub-bands of primary interest were in the regions 1000 ± 125 Hz and 2500 ± 250 Hz of the frequency axis. These are the areas in which the wavelet packet-based approximation of the critical bandwidth changes from a lower value (finer resolution) to higher values (coarser resolution). Secondly, taking into account the approximate character of the ERB and the approximate estimation of the critical bandwidth, Siafarikas et al. performed an objective study of the various (redundant) transforms that could cover these areas.

In brief, after the objective evaluation of 225 candidate wavelet packet trees, Siafarikas et al. (2005) reported that the resolutions and overlapping areas shown in Table 4.4 provide the most advantageous speaker verification performance among all tested candidates. Each candidate wavelet packet decomposition tree was evaluated in five subsequent experiments and the results averaged, for reducing the chance that any variability in the training or in the pattern matching stage, would affect the choice of sub-bands. Eventually, the wavelet coefficient vectors for the selected decomposition tree, consisting of $B = 73$ sub-bands are represented by the subset $S_{OVL,8kHz} = \{[\mathbf{W}_7^0, \mathbf{W}_7^{31}], [\mathbf{W}_6^{14}, \mathbf{W}_6^{23}], [\mathbf{W}_5^{12}, \mathbf{W}_5^{15}], [\mathbf{W}_6^{32}, \mathbf{W}_6^{41}], [\mathbf{W}_5^{19}, \mathbf{W}_5^{23}], [\mathbf{W}_6^{48}, \mathbf{W}_6^{55}], [\mathbf{W}_5^{28}, \mathbf{W}_5^{31}]\}$ shown in Fig. 4.7. The areas of overlapping are: the

---

[37] Definition of the equal error rate error is offered in Sect. 7.1

64
Contemporary Methods for Speech Parameterization

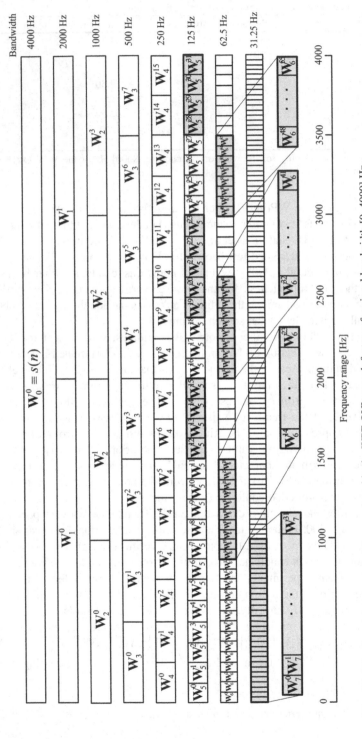

Fig. 4.7 Wavelet packet decomposition employed in the WPF-OVL speech features for signal bandwidth [0, 4000] Hz

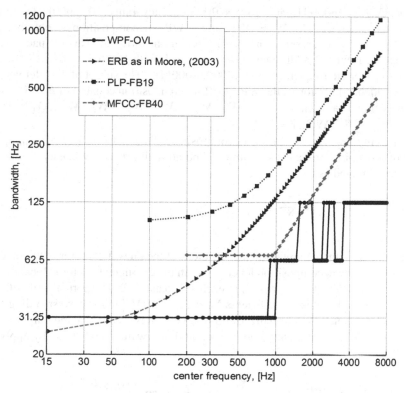

Fig. 4.8 The frequency warping for the WPF-OVL filter-bank and the ERB values (Moore 2003) computed for these center frequencies in comparison to the MFCC-FB40 and the PLP-FB19 filter-banks

frequency range [875, 1000] Hz covered with resolutions 31.25 Hz and 62.5 Hz; and the frequency range [2375, 2625] Hz covered with resolutions 62.5 Hz and 125 Hz. In Fig. 4.8, on the plot for WPF-OVL, these areas of overlapping are presented with two different values of the filter bandwidth for the same value of the center frequency. The parameters corresponding to this wavelet packet tree are as follows: wavelet packet decomposition of level $D = 7$, and the redundancy factors are $M_6^{16} = -2$, $M_5^{20} = -1, M_6^{39} = 2$. All other redundancy factors are equal to zero ($M_j^n = 0$). Thus, the overlapping parameter is $M = 5$.

It is worth mentioning, that although the objective evaluation of 225 wavelet packet trees reported in Siafarikas et al. (2005) is the largest objective evaluation of this kind reported in the literature, the wavelet packet decomposition shown in Fig. 4.7 is almost certainly only a local minima of the equal error rate function. The evaluated 225 wavelet packet trees are only a small fraction of the feasible trees offered by the ODWPT, and only an exhaustive search of all feasible trees could guarantee reaching the global minima for the particular task.

Like the other wavelet packet-based features discussed here, let us also consider a wideband version of the WPF-OVL filter-bank, which is designed for sampling frequency $f_s = 16000$ Hz and covers the frequency range [0, 8000] Hz. In the wideband version of the WPF-OVL, the DWPT decomposition is obtained for $D = 8$, in order to preserve the resolution of the original wavelet packet tree unchanged. Next, 32 additional sub-bands, each with bandwidth of 125 Hz, are added to cover the frequency range [4000, 8000] Hz. The subset of wavelet vectors for the wideband version of the WPF-OVL decomposition consists of $B = 105$ sub-bands defined as $S_{OVL,16kHz} = \{[\mathbf{W}_8^0, \mathbf{W}_8^{31}], [\mathbf{W}_7^{14}, \mathbf{W}_7^{23}], [\mathbf{W}_6^{12}, \mathbf{W}_6^{15}], [\mathbf{W}_7^{32}, \mathbf{W}_7^{41}],$ $[\mathbf{W}_6^{19}, \mathbf{W}_6^{23}], [\mathbf{W}_7^{48}, \mathbf{W}_7^{55}], [\mathbf{W}_6^{28}, \mathbf{W}_6^{63}]\}$.

Following the wavelet packet decomposition, $S_{OVL} \in \{S_{OVL,8kHz}, S_{OVL,16kHz}\}$, a total of B disjoint sub-bands are formed. The normalized energy in each frequency band is computed as:

$$E_p = \frac{1}{N_p} \sum_{m=1}^{N_p} \left[\mathbf{W}_j^n(m) \right]^2, \mathbf{W}_j^n \in S_{OVL}, p = 1, 2, ..., B, \qquad (4.14)$$

where $\mathbf{W}_j^n(m)$ is the mth individual coefficient of the ODWPT vector at the specific node $\mathbf{W}_j^n$, j is the decomposition level at which the coefficient vector is located, p is the sub-band index. The energy E_p in each sub-band is further normalized with the number of wavelet packet coefficients $N_p = N/2^j + \left| M_j^p \right|$ in the corresponding sub-band. Subsequently, the normalized sub-band energies obtained at the output of the filter-bank are logarithmically compressed and afterward decorrelated by applying the DCT:

$$WPFOVL(r) = \sqrt{\frac{2}{B}} \sum_{p=0}^{B-1} \log_{10}(E_{p+1}) \cos\left(\frac{r(p + 0.5)\pi}{B}\right),$$

$$r = 0, 1, ..., R - 1, \qquad (4.15)$$

where B is the total number of frequency sub-bands, and $R \leq B$ is the number of unique WPF-OVL cepstral coefficients. For larger R, the values of the WPF-OVL with index $r \geq B$ mirror these of the first B coefficients. The scale factor $\sqrt{2/B}$ makes the DCT matrix orthogonal, and for the same reason the cepstral coefficient with index $r = 0$ is next multiplied by the term $\sqrt{2}/2$. According to the definition of the WPF-OVL filter-banks above, one may have $R \leq 73$ unique coefficients for $f_s = 8000$ Hz and bandwidth [125, 4000] Hz, and $R \leq 105$ for $f_s = 16000$ Hz and bandwidth [0, 8000] Hz. Finally, the actual number of WPF-OVL parameters that are computed depends on the particular application.

4.5   The WPF-ACE

Similar to the speech parameterization methods described in the previous sections, cochlear implants implement sub-band processing of audio signals. In order to transmit the content of each sub-band to the nerve fibers of the hearing system, they rely on array of electrodes inserted in the cochlea.

In a number of studies involving patients with cochlear implants, pitch perception was pointed out as a particularly important aspect for the reliable comprehension of speech. In the present-day cochlear implants, there are limited possibilities for improving the perception of *place pitch* due to the limited number of electrodes which is implanted. Accounting for this limitation, Nogueira et al. (2006), investigated ways to improve the temporal resolution of pitch in cochlear implants built on the Advanced Combinational Encoder (ACE) strategy. Specifically, in order to allow enhanced perception of the *temporal pitch*, Nogueira et al. proposed a filter-bank and wavelet packet-based signal decomposition scheme, which replaces the DFT-based analysis in the standard ACE strategy. In that study, Nogueira et al. experimented with various wavelet functions and investigated their applicability for optimizing the temporal resolution of signal analysis. In brief, three different mother wavelet functions were evaluated for the implementation of the wavelet packet decomposition: (i) the Haar wavelet, (ii) the Daubechies wavelet of order 3, and (iii) a mixed wavelet packet tree based on the Symlets family of wavelet functions. Here, we focus on the mixed wavelet packet decomposition tree based on the Symlets family, as in a comparative subjective evaluation with patients, this implementation was reported as the most advantageous in terms of intelligibility of speech.

The specific model of cochlear implant, which Nogueira et al. investigated, had 22 electrodes. Under this strict restriction for the number of electrodes, and therefore for the number of stimulated sub-bands, Nogueira et al. selected a filter-bank design, which closely approximates the Bark scale and divides the frequency range [0, 8000] Hz in 21 intervals. However, the first filter (Fig. 4.9), which corresponds to the frequency range [0, 125] Hz was not used in the subsequent signal processing steps as it is believed that this sub-band contributes little to comprehension of speech.

Furthermore, each of the two filters with bandwidth of 750 Hz, which cover the frequency range [3500, 5000] Hz, are composite and their output is obtained by combining three groups of wavelet transform coefficients. These groups are obtained at level 5 of the wavelet packet decomposition (refer to Fig. 4.10), where each vector of wavelet coefficients corresponds to a sub-band of the original signal with bandwidth of 250 Hz. In this manner, the effective size of the filter-bank of Nogueira et al. is equivalent to 20 filters. In the work of Nogueira et al., the filter-bank outputs are next processed for selecting the N active electrodes (typically 8–12) out of the total $M = 20$ electrodes, which are then encoded in a "NofM" mapping of audio to excitation stimuli, which are next transmitted by the electrodes to the nerve fibers in the cochlea.

Here, we make use of the abovementioned filter-bank outputs in the context of speech parameterization. Adhering to the concept of short-time cepstral analysis of speech, we firstly compress logarithmically the magnitude of the filter-bank outputs and then de-correlate their values by applying the DCT. In the remaining sections, we refer to this speech parameterization scheme – that is, wavelet packet decomposition with mixed wavelet packet tree based on the Symlets family and a filter-bank approximating the Bark scale as in Nogueira et al. (2006) – as to wavelet packet features for the ACE strategy, or shortly WPF-ACE.

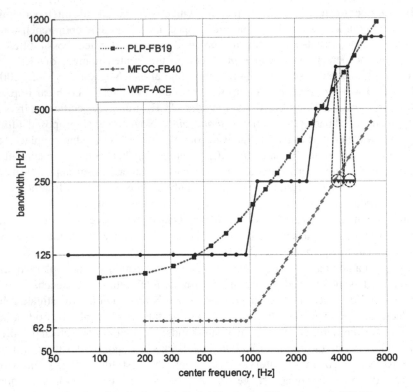

Fig. 4.9 Approximation of the Bark scale with the filter-bank of Nogueira et al. (2006) in comparison to the filter-banks used in the PLP-FB19 and in the MFCC-FB40 speech parameterizations

The computation of the WPF-ACE speech parameters is summarized as follows. Let us denote with $x(n)$ the discrete-time speech signal, sampled with frequency f_s, where n is the discrete time index. Let us consider that the signal $x(n)$ has been pre-processed as explained in Sect. 2.2. The outcome of this pre-processing is the pre-emphasized speech signal $s(n)$, which has been weighted with a rectangular window with length of N samples, where $N = 2^J$ is an exact power of two for some positive integer J. By applying the DWPT, the speech signal, $s(n)$, is partitioned as

$$\mathbf{W}_j^{2n}(k) = \sum_{i=0}^{L-1} a_{n,i} \mathbf{W}_{j-1}^n \left[(2k + 1 - i) \bmod \left(N/2^{j-1} \right) \right], \qquad (4.16)$$

$$\mathbf{W}_j^{2n+1}(k) = \sum_{i=0}^{L-1} b_{n,i} \mathbf{W}_{j-1}^n \left[(2k + 1 - i) \bmod \left(N/2^{j-1} \right) \right], \qquad (4.17)$$

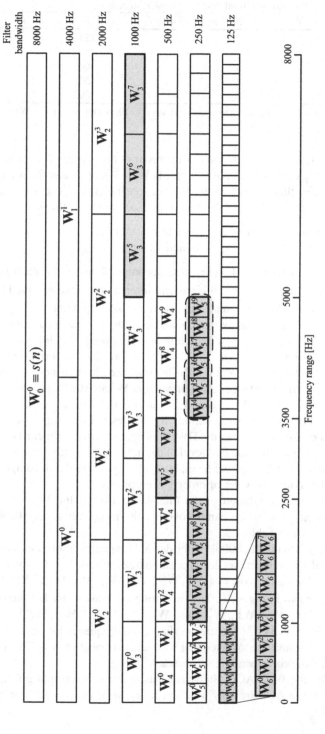

Fig. 4.10 The wavelet packet decomposition employed in the WPF-ACE speech features for signal bandwidth [0, 8000] Hz

Table 4.5 Frequency division used in the WPF-ACE filter-bank

Frequency range [Hz]	Number of sub-bands	Frequency resolution [Hz]
[0, 1000]	8	125
[1000, 2500]	6	250
[2500, 3500]	2	500
†[3500, 5000][a]	2	750
[5000, 8000][a]	3	1000

[a] These sub-bands exist only for the wideband version [0, 8000] Hz of the WPF-ACE speech features

where $k = 0, 1, \ldots, N/2^j - 1$, and L is the order of the wavelet function. The parent signal on the top of the decomposition tree is $\mathbf{W}_0^0(n) \equiv s(n)$, and the wavelet filter h_i and the scaling filter g_i represented by the coefficients $a_{n,i}$ and $b_{n,i}$ are applied depending on their position in the decomposition tree, as follows:

(i) For the *even* values of n, $a_{n,i} \equiv g_i$ and $b_{n,i} \equiv h_i$
(ii) For the *odd* values of n, $a_{n,i} \equiv h_i$ and $b_{n,i} \equiv g_i$

By going from level $j - 1$ to the next decomposition level j, each *parent node* $\mathbf{W}_{j-1}^{n'}$ of the decomposition tree is circularly filtered and down-sampled twice: once with the wavelet filter $\{h_l\}$ and once with the scaling filter $\{g_l\}$, yielding two *children nodes* $\mathbf{W}_j^n$, indexed by $n = 2n'$ and $n = 2n' + 1$. Each partition, $\mathbf{W}_j^n$, of the original signal $\mathbf{W}_0^0(n)$, obtained in this way contains the information for the frequency interval $[n/2^{j+1}, (n+1)/2^{j+1}]$ and provides information associated with the time interval $[2^j k, 2^j (k + 1)]$. Once all the children nodes of interest are computed, the corresponding coefficient vectors $\mathbf{W}_j^n$ are stacked together to form an orthogonal subset $S = \{\mathbf{W}_j^n : j = 0, 1, \ldots, J, n = 0, 1, \ldots, 2^j - 1\}$, which corresponds to the frequency division defined by the desired decomposition tree. In the computation of the WPF-ACE, the DWPT is implemented by means of the Symlet wavelet filter of order 6 at the first level of decomposition, with its impulse response effectively halved at the second level, etc., until reaching the decomposition level six, where the Symlet wavelet of order 1 is employed.

In the original formulation of the filter-bank used in the WPF-ACE, the frequency division with 20 sub-bands covering the frequency range [0, 8000] Hz is defined as shown in Table 4.5.

For signal bandwidth [0, 8000] Hz the desired frequency resolution is implemented with the use of a wavelet packet decomposition tree of depth $D = 6$, which leads to the orthogonal subset $S_{ACE,16kHz} = \{[\mathbf{W}_6^0, \mathbf{W}_6^7], [\mathbf{W}_5^4, \mathbf{W}_5^9], [\mathbf{W}_4^5, \mathbf{W}_4^6], [\mathbf{W}_5^{14 \div 16}, \mathbf{W}_5^{17 \div 19}],$ $[\mathbf{W}_3^4, \mathbf{W}_3^7]\}$, shown in Fig. 4.10. The two groups of wavelet packet transform coefficients $\mathbf{W}_5^{14 \div 16} = \{\mathbf{W}_5^{14}, \mathbf{W}_5^{15}, \mathbf{W}_5^{16}\}$ and $\mathbf{W}_5^{17 \div 19} = \{\mathbf{W}_5^{17}, \mathbf{W}_5^{18}, \mathbf{W}_5^{19}\}$, which define the interval $[\mathbf{W}_5^{14 \div 16}, \mathbf{W}_5^{17 \div 19}]$, corresponding to the frequency range [3500, 5000] Hz, are obtained after stacking together their wavelet coefficient vectors. This is equivalent to aggregating the corresponding sub-bands.

Originally, the WPF-ACE filter-bank was defined for sampling frequency $f_s = 16000$ Hz and the ACE strategy uses signal segments of $N = 128$ samples, which corresponds to frame size of 8 ms. However, for the purpose of fair comparison with the other speech parameterization schemes considered in this book, we consider frame size of either 16 ms ($N = 256$ samples) or 32 ms ($N = 512$ samples), depending on the application setup.

In addition, let us also define a narrowband version of the WPF-ACE for sampling rate $f_s = 8000$ Hz, where the frame size of 32 ms corresponds to $N = 256$. Since in this case the signal content is restricted to the frequency range [0, 4000] Hz, the last three sub-bands with bandwidth 1000 Hz are discarded from the wavelet packet tree (the last row in Table 4.5). The same holds for the second of the two sub-bands with bandwidth of 750 Hz (denoted with the symbol † in Table 4.5), and the bandwidth of the first one is reduced from 750 to 500 Hz. As for $f_s = 8000$, the signal $s(n)$ has only half of the frequency range and half the number of samples (for the same frame size) one needs to decrease the depth of the wavelet packet tree to $D = 5$ in order to preserve the original resolution of the filter-bank as defined in Nogueira et al. (2006). Therefore, for this narrowband version of the WPF-ACE parameters, we make use of the orthogonal subset $S_{ACE,8kHz} = \{[\mathbf{W}_6^0, \mathbf{W}_6^7], [\mathbf{W}_5^4, \mathbf{W}_5^9], [\mathbf{W}_4^5, \mathbf{W}_4^6], \mathbf{W}_5^{14 \div 15}\}$, where the composite term $\mathbf{W}_5^{14 \div 15}$ is defined as a group of two wavelet packet transform vectors $\mathbf{W}_5^{14 \div 15} = \{\mathbf{W}_5^{14}, \mathbf{W}_5^{15}\}$. The bandwidth defined by the composite vector $\mathbf{W}_5^{14 \div 15}$ is equal to the sum of the bandwidths of the two sub-bands, which constitute it.

Following the wavelet packet decomposition, $S_{ACE} \in \{S_{ACE,8kHz}, S_{ACE,16kHz}\}$, in total B disjoint sub-bands are formed. The energy E_p in each sub-band p is calculated as follows:

$$E_p = \frac{1}{N_p} \sum_{m=1}^{N_p} \left[\mathbf{W}_j^n(m)\right]^2, \mathbf{W}_j^n \in S_{ACE}, p = 1, 2, ..., B, \quad (4.18)$$

where $\mathbf{W}_j^n(m)$ is the mth individual coefficient of the DWPT vector at the specific node $\mathbf{W}_j^n$, j is the depth level at which the coefficient vector is located, and p is the sub-band index. Here, the energy E_p in each sub-band is normalized with the number of wavelet packet coefficients N_p in the corresponding sub-band.

Subsequently, the normalized sub-band energies obtained at the output of the filter-bank (4.18) are compressed logarithmically and are decorrelated by applying the DCT:

$$WPFACE(r) = \sqrt{\frac{2}{B}} \sum_{p=0}^{B-1} \log_{10}(E_{p+1}) \cos\left(\frac{r(p + 0.5)\pi}{B}\right),$$
$$r = 0, 1, ..., R - 1. \quad (4.19)$$

Here, B is the total number of frequency sub-bands and $R \leq B$ is the number of unique WPF-ACE cepstral coefficients to be computed. For larger R, the values of the WPF-ACE with index $r \geq B$ mirror these of the first B coefficients. The scale factor $\sqrt{2/B}$ makes the DCT matrix orthogonal, and for the same reason the cepstral coefficient with index $r = 0$ is next multiplied by the term $\sqrt{2}/2$. According to the definition of the WPF-ACE filter-bank above, one may have $R \leq$ 20 unique coefficients for $f_s = 16000$ Hz and signal bandwidth [0, 8000] Hz, and $R \leq 16$ for $f_s = 8000$ Hz and signal bandwidth [0, 4000] Hz. The actual number of WPF-ACE cepstral coefficients to be computed depends on the application setup.

4.6 Comparison Among the Various DWPT-Based Speech Features

This section offers a brief account of the main differences among the five DWPT-based speech parameterizations discussed in Sect. 4, and their similarities and dissimilarities with the DFT-based methods considered in Sect. 3.

First of all, it should be said that all speech parameterizations discussed in Sect. 4 share common speech pre-processing steps, which consist of signal pre-emphasis and framing with rectangular window (details in Sect. 2.2). The use of rectangular window is in contrast to the DFT-based speech parameterizations, where appropriate weighting (most often with the Hamming or Hann function) is required for avoiding abrupt changes of the signal amplitude at the speech frame boundaries, and for reducing the spectral leakage in the frequency domain. Furthermore, all DWPT-based speech features discussed here follow uniform computational algorithm (refer to the block diagram shown in Fig. 2.1, Sect. 2.1), similar to the one used in the various MFCC implementations. The most significant difference is the use of DWPT instead of the DFT at the time-frequency decomposition step.

The main differences among the various DWPT-based speech parameterizations discussed here are in the choice of basis wavelet function for the time–frequency decomposition and the design of the wavelet packet tree. While in the DFT the basis functions are fixed to sin(.) and cos(.), the DWPT has the flexibility to make use of the variety of existing wavelet functions or to take advantage of a purposely designed wavelet function for the particular type of signal. For instance, the WPF-SBC make use of the Daubechies wavelet of order 32, the WPF-FD rely on the Daubechies wavelet of order 12, the WPF-OVL and WPF-OBJ are implemented by means of the Battle-Lemarié wavelet of order 5, and the WPF-ACE make use of the Symlet of order 6. These wavelet functions offer different trade-offs between frequency and time resolution and their choice depends on the properties of the signal to be analyzed and on the objectives of the particular signal processing task.

Figure 4.11 shows the magnitude response in dB for the four wavelet functions of interest: (i) Daubechies wavelet of order 12 (dashed line); (ii) Daubechies wavelet of order 32 (dotted line); (iii) Symlet wavelet of order 6 (dash-dotted line); and (iv) Battle-Lemarié wavelet of order 5 (solid line). The area enclosed with the dotted rectangle in the figure is magnified on the back image so that the slopes of the magnitude response around the area of -3 dB are seen.

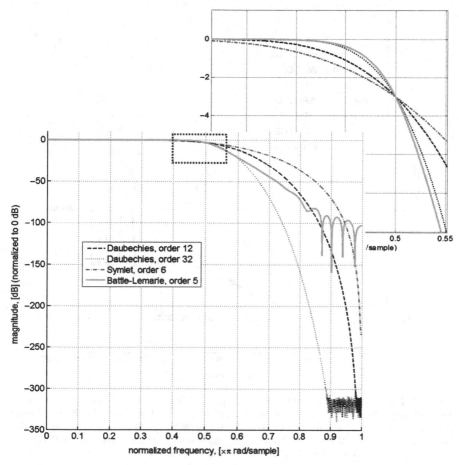

Fig. 4.11 Magnitude response in dB for four wavelet functions: (i) Daubechies wavelet of order 12 (*dashed line*); (ii) Daubechies wavelet of order 32 (*dotted line*); (iii) Symlet wavelet of order 6 (*dash-dotted line*); and (iv) Battle-Lemarié wavelet of order 5 (*solid line*)

As the figure shows, in the magnified area, the magnitude response of the Battle-Lemarié wavelet of order 5 is closer to the perfect low-pass filter when compared to those of the other wavelet functions. Another advantage of the Battle-Lemarié wavelet is that it has a linear phase response, while this is not the case for the other wavelet functions. However, the Battle-Lemarié wavelet has the lowest suppression in the stop-band, which is nearly three times worse than that of the Symlet wavelet of order 6 and nearly four times worse than that of the Daubechies wavelets of order 12 and 32.

These properties of the Battle-Lemarié wavelet make it relatively more advantageous for the analysis of speech signals at least from theoretical perspective. Starting from the same considerations, Siafarikas et al. (2007) investigated the appropriateness of five wavelet functions for the needs of the speaker verification task. All wavelet functions were evaluated in a common experimental setup and it was

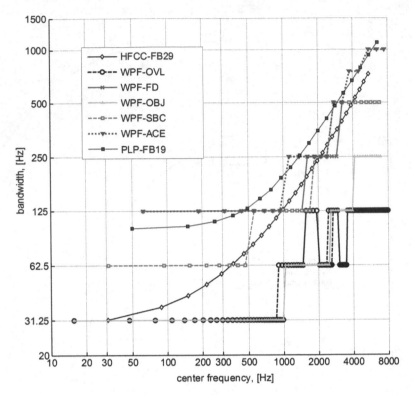

Fig. 4.12 Frequency warping used in various DFT- and DWPT-based speech parameterizations. The Bark scale (PLP-FB19) – *solid line and marker* "■"; the ERB scale (HFCC-FB29) of Moore and Glasberg (1983), – *solid line and marker* "◇"; the WPF-SBC – *dashed line with marker* "□"; WPF-FD – *solid line with marker* "x"; WPF-OBJ – *light solid line with marker* "▲"; WPF-OVL – *dashed line with marker* "o"; WPF-ACE – *dotted line with marker* "▼"

reported that the Battle-Lemarié wavelet of order 5 leads to speech features, which offer the highest speaker verification accuracy.

Another major difference among the WPF-SBC, WPF-FD, WPF-OBJ, WPF-OVL, and WPF-ACE is the frequency warping, that is, the filter-bank design. As explained in Sect. 4, these DWPT-based speech parameterizations employ different wavelet packet decomposition trees. This is equivalent to using filter-banks with different number of filters and different spacing of the center frequencies of these filters.

Figure 4.12 shows the frequency warping used in the five DWPT-based speech parameterizations in comparison to the Bark-scale frequency warping used in the PLP-FB19 speech features, and to the ERB frequency warping as defined in Moore and Glasberg (1983), which is used in the HFCC-FB29 speech features. As Fig. 4.12 shows, the WPF-ACE filter-bank closely follows the Bark scale frequency warping. The filter-banks used in the WPF-SBC and WPF-FD, which approximate the Mel scale follow closely the ERB scale of Moore and Glasberg (1983), used in the HFCC-FB29, except for the filters with central frequency below 500 Hz, where the ERB concept suggests a smaller filter bandwidth. Finally, the WPF-OBJ and WPF-OVL use frequency warping which is proportional to the ERB concept but the width of the individual sub-bands is approximately two to three times smaller than the ERB values for most of the range. The last is due to the larger number of sub-bands in the WPF-OBJ and WPF-OVL frequency decomposition (two to three times more sub-bands when compared to the number of filters in the HFCC-FB29 filter-bank). On the other hand, the number of sub-bands in the WPF-ACE is commensurate with the number of filters in the PLP-FB19 and the number of sub-bands in the WPF-FD and WPF-SBC is commensurate with the number of filters in the MFCC-FB24 (not shown in the figure).

In consequence of the larger number of sub-bands, the WPF-OBJ and WPF-OVL speech parameterizations allow the computation of a larger number of unique cepstral coefficients when compared to the other DFT- and DWPT-based speech parameterizations. For instance, one may compute up to 100 and 105 cepstral coefficients, for the WPF-OBJ and WPF-OVL speech parameterizations, respectively. In this manner, the WPF-OBJ and WPF-OVL offer a high-resolution cepstral analysis, that is, the speech cepstrum is sampled by a larger number of cepstral coefficients, which is beneficial for the needs of the speaker recognition tasks as this allows more speaker-specific details to be revealed and exploited for distinguishing among speakers with similar voice traits.

The number of unique cepstral coefficients that can be computed for the WPF-ACE, WPF-FD, and WPF-SBC are 20, 24, and 32, respectively, which are comparable to the number of unique cepstral coefficients (between 19 and 40), which can be computed in the various MFCC implementations. The larger number of cepstral coefficients offers the opportunity to design larger dimensionality feature vectors, which capture the finer details in the speech signal (both through the cepstral coefficients with large indexes and through the higher resolution of the cepstrum). Capturing the finer variability is expected to benefit the speaker recognition tasks, but not the speech recognition and other tasks, where the fine-grain variability is not desired. Experimental evaluation of these speculations is offered in Sects. 6 and 7.

5 Evaluation on the Monophone Recognition Task

Section 5 offers a comprehensive evaluation of twelve speech parameterizations on the monophone recognition task.[38] Specifically, the monophone recognition accuracy of seven DFT-based speech feature sets (PLP-FB18, LFCC-FB40, MFCC-FB23,

---

[38] The monophone recognition task as understood here is the phone-to-grapheme conversion of speech without considering the context in which the specific phone appears or the linguistic dimension of the particular language. Here, we use the term *monophone* for clarity and contrast to the *triphone* recognition problem.

MFCC-FB40, HFCC-FB23, HFCC-FB28, and HFCC-FB40) outlined in Sect. 3 and the five DWPT-based features (WPF-SBC, WPF-FD, WPF-OBJ, WPF-OVL, and WPF-ACE) outlined in Sect. 4, was investigated in a common experimental setup. For that purpose, we made use of the well-known open-source HMM ToolKit (HTK) (Young et al. 2006) and the widely used TIMIT speech recognition database (Garofolo 1998), which offers a phone-level annotations and a well-understood experimental protocol. This experimental setup shall allow the reader to carry out a direct comparison with related studies, which investigate other speech parameterizations on the monophone recognition task.

5.1   The Monophone Recognition Task

Due to the relatively small computational and memory demands when compared to other speech processing tasks, the monophone recognition and the isolated mono-syllabic word recognition were in the center of research interest during the 1970s and 1980s. In this section, we focus on the monophone recognition task as it has proved to be quite a challenging problem, and because in the last four decades it was established as the common benchmark for the evaluation of speech technology innovations and, in particular, new speech parameterization methods.

As in many other speech processing applications, the accuracy obtained on the monophone recognition task is language dependent, and thus, it is quite difficult to make parallels between results obtained for different languages. In this section, due to the use of the TIMIT database, we only report monophone recognition results for English language. It is quite likely that the ranking of the speech parameterization will change for another language, as the phonetic structure might differ significantly among the major language groups. Furthermore, as we will see in the next sections, the ranking of the speech feature sets depends on the particular application, and as the results from numerous studies show, also on the particular experimental setup. Therefore, the ranking order of speech features reported in this section should not be considered absolute,[39] but is valid for the monophone recognition task, English language, HMM-based modeling, and the particular experimental setup.

The traditional measures of accuracy in the evaluations of technology on the monophone recognition task are the *monophone recognition accuracy* and the *average monophone classification rate* in percentages. The monophone recognition accuracy is the most representative measure of accuracy as it accounts for all

---

[39] The purpose of this book is not to promote a particular speech parameterization method but to caution the reader that the choice of speech parameterization need to be coherent with the requirements of the particular application, choice of modeling method, language, operational setup, presence of interferences and noise, etc. These arguments apply not only for the ranking results presented in Sects. 5–7, but also to all studies reporting on the advantage of particular speech parameterization scheme.

possible types of errors occurring during real-life operation of a system. It is defined (Young et al. 2006) as

$$RA_{mph} = \frac{N_{mph} - (D_{mph} + S_{mph} + I_{mph})}{N_{mph}} \cdot 100\%. \tag{5.1}$$

Here, N_{mph} is the total number of monophone instances in the test dataset, D_{mph} is the number of *deletion errors* (monophones that were omitted in the automatically obtained phone sequence when compared to the reference ground-truth transcriptions), S_{mph} is the number of *substitution errors* (monophones that were misrecognized as another monophone), and I_{mph} is the number of *insertion errors* (monophones that appear in the obtained phone sequence but do not exist in the reference ground-truth transcription of the spoken utterance). As follows from Eq. 5.1, the value of the monophone recognition accuracy, RA_{mph}, is not bound in the range [0, 100] % and for a really ineffective recognizer RA_{mph} can take negative values. This happens when the sum in the brackets in the numerator of (5.1) exceeds the total number of instances N_{mph}. Hence, although it is the more representative measure of recognition accuracy, the monophone recognition accuracy is not an intuitive measure. For that reason, the percentage of correctly classified monophones, defined as

$$ACR_{mph} = \frac{N_{mph} - S_{mph}}{N_{mph}} \cdot 100\% \tag{5.2}$$

is often provided as an additional measure of classification accuracy. As seen from (5.2) ACR_{mph} does not account for the insertion and deletion errors, and its values are bounded in the range [0, 100] %. In the Sect. 5.3 these two accuracy measures are jointly employed as indicators of the monophone recognition performance for the speech parameterizations of interest.

Furthermore, since in the subsequent Sects. 6 and 7 all experimental results are presented graphically in terms of error rates, we will also make use of the reciprocal values of RA_{mph} and ACR_{mph}. This is to avoid misinterpretation of the graphical representation of the ranking of speech features on the monophone recognition task. The reciprocal quantities of RA_{mph} and ACR_{mph} have the meaning of error rates and are referred to as *monophone recognition error* (RE) and *average monophone classification error* (ACE).

For the monophone recognition error, we have $RE_{mph} = 100 - RA_{mph}$, which is equivalent to

$$RE_{mph} = \frac{D_{mph} + S_{mph} + I_{mph}}{N_{mph}} \cdot 100\%. \tag{5.3}$$

In the same manner, for the average monophone classification error, we have $ACE_{mph} = 100 - ACR_{mph}$, or

$$\text{ACE}_{mph} = \frac{S_{mph}}{N_{mph}} \cdot 100\%. \tag{5.4}$$

As seen from (5.3) and (5.4), the monophone recognition error, RE_{mph}, and the average monophone classification error, ACE_{mph}, are not independent variables. However, their joint use offers the opportunity, the results for all speech parameterization methods to be mapped on a common plane, defined by RE_{mph} and ACE_{mph}, which we consider more illustrative than using only the one-dimensional ranking in terms of RE_{mph}.

5.2 Experimental Protocol

During the past decades of research efforts on the monophone recognition task, the HMM ToolKit (Young et al. 2006) and the TIMIT speech recognition database (Garofolo 1998) were established as the standard test-bed for benchmarking different modeling techniques or for promoting innovations, among which are new speech parameterizations. For the reason of compatibility with previous research on the topic, we follow the standard experimental protocol, which makes use of the recommended train and test splits of the TIMIT dataset.

In brief, TIMIT consists of studio-quality speech recordings of phonetically balanced sentences from 630 American-English speakers representing eight dialects. For each speaker, there are ten speech recordings, sampled at rate 16000 Hz with 16-bit precision per speech sample. The database is divided into balanced training and testing sets of 462 and 168 speakers, respectively. The total number of monophone instances in the TIMIT test subset is 53875.

The speech parameterization of interest are the seven DFT-based speech feature sets (PLP-FB18, LFCC-FB40, MFCC-FB23, MFCC-FB40, HFCC-FB23, HFCC-FB28, and HFCC-FB40) outlined in Sect. 3 and the five DWPT-based features (WPF-SBC, WPF-FD, WPF-OBJ, WPF-OVL, and WPF-ACE) outlined in Sect. 4. A common experimental protocol and pre-processing of the speech data were followed. In particular, we considered speech frame size of 32 ms and skip step of 10 ms between two subsequent frames. This window size was imposed by the requirement of DWPT-based speech parameterizations for number of samples which is an exact power of two. For all speech parameterizations, we computed only the first 13 cepstral coefficients and final feature vector was obtained by appending their first and second time derivatives computed for a window of ±3 frames. In this manner, every speech frame is represented by a 39-dimentional feature vector. No automatic gain control and variance normalization were applied on the feature vectors.

The 39 symbol ARPAbet[40] phone set {/aa/, /ae/, /ah/, /ao/, /aw/, /ay/, /b/, /ch/, /d/, /dh/, /eh/, /er/, /ey/, /f/, /g/, /hh/, /ih/, /iy/, /jh/, /k/, /l/, /m/, /n/, /ng/, /ow/, /oy/, /p/,

---

[40] ARPAbet phonetic transcription, http://en.wikipedia.org/wiki/Arpabet

Table 5.1 Adaptation of the frequency range for the filter-banks and sub-bands of the evaluated speech parameterizations

Speech parameterization	Discarded filters/sub-bands	Designation
HTK MFCC-FB26	The first and the last two filters	HTK MFCC-FB23
HFCC-FB29	The first filter	HFCC-FB28
LFCC-FB48	The first filter and the last seven filters	LFCC-FB40
PLP-FB19	The last filter	PLP-FB18
WPF-SBC	The first two and the last two sub-bands	WPF-SBC
WPF-FD	The first and the last sub-bands	WPF-FD
WPF-OBJ 125	The first four and the last eight	WPF-OBJ 125
WPF-OBJ 250	The first four and the last four	WPF-OBJ 250
WPF-OVL	The first four and the last four	WPF-OVL
WPF-ACE	The last sub-band	WPF-ACE

/r/, /s/, /sh/, /t/, /th/, /uh/, /uw/, /v/, /w/, /y/, /z/, /zh/} plus silence, /sil/, were modeled with three-state HMMs implemented with the HTK (Young et al. 2006). Each of the three states is implemented with a 12-component Gaussian mixture model (GMM), which was found out the optimal for the HTK MFCC-FB24 feature set, and the model for silence has 24-component GMM. All the 40 three-state HMMs have self-loops but no skip transitions over the states. The HMMs were trained with diagonal covariances by using a splitting procedure, where the training starts with one mixture and is reestimated after each split. The training process was terminated with error stopping criterion, which resulted in different number of iterations (between 40 and 60) for each speech parameterization.

5.3 Comparative Evaluation on the Monophone Recognition Task

All speech parameterizations of interest were processed in uniform manner as described in the Sect. 5.2. Furthermore, all speech parameterizations were adapted to the bandwidth of the MFCC-FB40 filter-bank [133, 6855] Hz by discarding the filters which reside outside this range. These changes are summarized in Table 5.1. In addition, we added the HFCC-FB23 speech features, which were computed for the same frequency range.

The HTK MFCC-FB23 speech parameters are considered the reference point (baseline), as they are the default speech parameterization used in the HTK speech recognizer.

As mentioned in Sect. 5.2, for the purpose of fair comparison, we made use of speech frame size of 32 ms for all speech features, including for the DFT-based ones. For this frame size, the baseline HTK MFCC-FB23 obtain monophone recognition accuracy $RA_{mph} = 63.1\%$, which is slightly lower than the one obtained in the same setup but for the more commonly used frame size of 25 ms, $RA_{mph} = 63.8\%$.

Table 5.2 American-English monophone recognition results for twelve speech parameterizations – averaged monophone classification rate and monophone recognition accuracy in percentages

American English (39 monophones + silence)	PLP-FB18	HFCC-FB40	MFCC-FB23	HFCC-FB28	MFCC-FB40	HFCC-FB23	WPF-FD	WPF-SBC	WPF-OBJ	LFCC-FB40	WPF-OVL	WPF-ACE
Average monophone classification rate [%] Eq. 5.2	**74.4**	**74.1**	74.0	73.8	73.9	73.6	72.1	71.6	71.0	70.6	69.6	64.9
Monophone recognition accuracy [%] Eq. 5.1	**62.9**	**63.1**	**63.1**	62.8	62.4	62.6	60.3	59.9	59.2	59.0	57.6	52.1

However, the small performance drop from $RA_{mph} = 63.8\%$ to $RA_{mph} = 63.1\%$ due to the chosen frame size of 32 ms is of secondary concern here as we mainly aim at comparing the relative ranking of the various speech parameterizations on the monophone recognition task, and do not aim at achieving the highest possible accuracy on the TIMIT database.

Table 5.2 shows the average monophone classification rate and the monophone recognition accuracy for the twelve speech parameterizations of interest. The ordering of the columns in the table corresponds to the ranking of these speech features on the monophone recognition task, with the left-hand column corresponding to the highest accuracy and the right-hand column corresponding to the lowest. As seen in Table 5.2, the PLP-FB18 speech features achieved the best average monophone classification rate and the second-best monophone recognition accuracy. In terms of average monophone classification rate, the PLP-FB18 are closely followed by the different implementations of the HFCC and MFCC speech parameters (HFCC-FB40, MFCC-FB23, HFCC-FB28, MFCC-FB40, HFCC-FB23).

Nearly the same ranking is obtained in terms of monophone recognition accuracy, with the only exception that the PLP-FB18 speech features show the second-best performance after the HFCC-FB40 and MFCC-FB23. Next, the LFCC-FB40 speech features, which are computed over the linear frequency scale, show significantly worse results than the PLP, MFCC, and HFCC speech parameters, which use frequency warping motivated by the properties of the human auditory system.

Furthermore, all the DWPT-based speech parameterizations, WPF-FD, WPF-SBC, WPF-OBJ, WPF-OVL, WPF-ACE, show significantly inferior average monophone classification rates and monophone recognition accuracy when compared to the PLP-FB18, MFCC, and HFCC speech features.

In order to investigate the statistical significance of the observed differences in the monophone recognition accuracy, we performed the matched pair statistical significance test MAPSSWE[41] applied on monophone level segments for

---

[41] MAPSSWE – The Matched Pairs Sentence Segment Word Error Test is part of the NIST Scoring Toolkit (SCTK) Version 2.4 (November 2009), available online at the NIST Web site: http://www.itl.nist.gov/iad/894.01/tools/

significance difference of 99.9% (threshold $p = 0.001$). The results of the pair wise evaluation of statistical significance are presented in Table 5.3. The cells with gray background and double-line boxes indicate pairs of results which are statically the same (labeled "same") or different with confidence level smaller than 99.9%. In both cases, the p-value is provided in brackets, and in the cases of statistical significance of the difference at level 95% and 99% we also show which of the two speech feature sets is better. Finally, for the pairs of speech parameterizations where the recognition accuracy results are different with confidence over 99.9%, we have omitted the p-value (in all cases $p < 0.001$), and show which of the two speech parameterizations is the better one. The statistical significance results for some of the speech parameterizations that show equivalent recognition accuracy are presented also in Fig. 5.1, but not all so as to keep the figure readable.

As Table 5.3 presents, the monophone recognition accuracy results for the PLP-FB18 are statistically no different from these for the MFCC and HFCC speech features (HFCC-FB40, MFCC-FB23, HFCC-FB28, MFCC-FB40, and HFCC-FB23). Statistical significance results for these are not shown in Fig. 5.1 to avoid overburdening of the illustration. Next, the results for the HFCC-FB40 are statistically different from these for MFCC-FB40 and HFCC-FB23, and the results for the MFCC-FB23 are different from those of MFCC-FB40, which gives an advantage of the HFCC-FB40 and MFCC-FB23 over the other MFCC and HFCC speech parameterizations. However, according to the statistical significance test, the use of PLP-FB18 offers equivalent recognition accuracy.

Next, the monophone recognition accuracy for the pairs WPF-FD and WPF-SBC, and the LFCC-FB40 and WPF-OBJ, turned out to be equivalent for statistical significance evaluated at confidence level 99.9%. Finally, as shown in Fig. 5.1, the WPF-OVL and the WPF-ACE demonstrated much worse monophone recognition accuracy when compared to the best-performing speech parameterizations.

The low performance of the WPF-OBJ and the WPF-OVL speech features was expected as they were purposely designed (and optimized in objective manner) for the text-independent speaker verification task. Thus, to some degree they were tuned up to emphasize the differences among speakers, and decrease the influence of the linguistic contents of speech, which is not beneficial to the monophone recognition task. In the same manner, the WPF-ACE filter-bank was tuned up for the needs of hearing-aids, and was not designed with the monophone recognition task in mind. In previous related studies, the accuracy for the LFCC-FB40 speech features was also shown inferior to the one for the MFCC-FB40, so their lower performance here was expected. However, when compared to the DFT-based speech parameterizations, the relatively lower performance of the WPF-SBC and WPF-FD is surprising, as in previous related work (Sarikaya and Hansen 2000; Farooq and Datta 2001) they were shown to outperform the MFCC-FB23.

We can explain the observed difference with the dissimilar evaluation setup, as Farooq and Datta (2001) reported the advantage of WPF-FD over the MFCC-FB24 on a small subset of the TIMIT database. This subset included only 15 out of the 39 monophones and only 37 test speakers instead of the entire test subset of 168 speakers. In the same manner, we can explain the difference with the results

Table 5.3 Results for the matched pair significance tests applied for the monophone recognition accuracy

	HFCC-FB40	MFCC-FB23	HFCC-FB28	MFCC-FB40	HFCC-FB23	WPF-FD	WPF-SBC	WPF-OBJ	LFCC-FB40	WPF-OVL	WPF-ACE
PLP-FB18	Same (0.395)	Same (0.834)	Same (0.294)	PLP-FB18 (0.003)	PLP-FB18 (0.020)	PLP-FB18	PLP-FB18	PLP-FB18	PLP-FB18	PLP-FB18	PLP-FB18
HFCC-FB40		Same (0.490)	HFCC-FB40 (0.016)	HFCC-FB40	HFCC-FB40	HFCC-FB40	HFCC-FB40	HFCC-FB40	HFCC-FB40	HFCC-FB40	HFCC-FB40
MFCC-FB23			Same (0.168)	MFCC-FB23	MFCC-FB23 (0.005)	MFCC-FB23	MFCC-FB23	MFCC-FB23	MFCC-FB23	MFCC-FB23	MFCC-FB23
HFCC-FB28				HFCC-FB28 (0.027)	Same (0.101)	HFCC-FB28	HFCC-FB28	HFCC-FB28	HFCC-FB28	HFCC-FB28	HFCC-FB28
MFCC-FB40					Same (0.465)	MFCC-FB40	MFCC-FB40	MFCC-FB40	MFCC-FB40	MFCC-FB40	MFCC-FB40
HFCC-FB23						HFCC-FB23	HFCC-FB23	HFCC-FB23	HFCC-FB23	HFCC-FB23	HFCC-FB23
WPF-FD							WPF-FD (0.006)	WPF-FD	WPF-FD	WPF-FD	WPF-FD
WPF-SBC								WPF-SBC	WPF-SBC	WPF-SBC	WPF-SBC
WPF-OBJ									Same (0.849)	WPF-OBJ	WPF-OBJ
LFCC-FB40										LFCC-FB40	LFCC-FB40
WPF-OVL											WPF-OVL

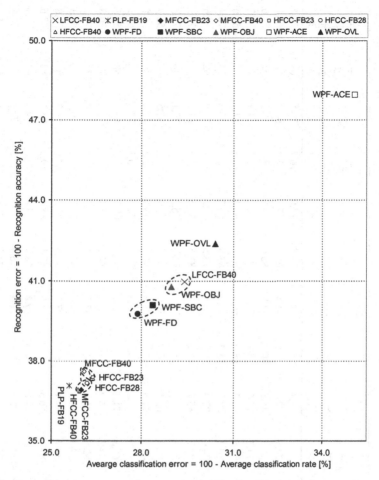

Fig. 5.1 Recognition error versus the average classification for the twelve speech parameterizations evaluated on the monophone recognition task

reported in Sarikaya and Hansen (2000), with the different evaluation setup. The WPF-SBC was reported advantageous over another implementation of the MFCC-FB24 speech features (Sarikaya et al. 1998), and on another database,[42] which is much smaller than TIMIT and does not cover the entire set of monophones.

Furthermore, as both studies (Farooq and Datta 2001; Sarikaya and Hansen 2000) offer details for the monophone classification accuracy on the level of monophone and phonetic category, in the following, we also detail the results presented in Table 5.2 to the level of individual consonants (Table 5.4) and individual vowels, diphthongs, and semivowels (Table 5.5).

---

[42] In Sarikaya and Hansen (2000), the authors made use of the SUSAS database (Speech Under Simulated and Actual Stress).

Table 5.4 American-English monophone recognition results for twelve speech parameterizations – percentage of correctly recognized monophones for the case of consonants

American English (39 monophones) consonants		PLP-FB18	HFCC-FB40	MFCC-FB23	HFCC-FB28	MFCC-FB40	HFCC-FB23	WPF-FD	WPF-SBC	WPF-OBJ	LFCC-FB40	WPF-OVL	WPF-ACE
Nasals	/m/	80.1	79.1	81.3	81.1	81.0	79.5	77.8	77.9	76.7	78.5	69.1	77.3
	/n/	73.0	68.9	72.5	72.5	72.0	71.4	68.7	67.1	71.0	68.3	53.3	68.1
	/ng/	84.6	84.1	84.3	85.5	80.9	84.9	80.6	75.8	78.8	79.1	65.3	81.6
Voiced plosives	/b/	75.1	74.8	74.2	75.5	75.7	75.5	71.3	74.2	76.1	74.1	74.7	74.2
	/d/	70.3	65.8	68.5	66.3	65.0	69.4	63.6	64.1	62.2	61.7	52.8	52.6
Unvoiced stops	/p/	81.1	84.0	80.4	83.2	83.2	80.7	85.0	83.2	79.9	82.3	74.0	76.4
	/t/	75.3	76.4	74.6	77.3	75.7	73.4	73.5	72.2	74.1	71.9	62.3	66.6
	/k/	83.8	85.3	85.5	85.2	85.1	85.4	81.5	82.5	82.9	81.4	68.0	79.9
Voiced stops	/g/	78.0	79.5	76.3	79.5	78.1	78.0	72.7	76.2	78.0	73.1	70.8	73.8
Voiced fricatives	/dh/	58.6	59.2	60.5	62.8	62.7	59.8	57.3	54.7	65.3	57.5	54.6	52.9
	/zh/	48.5	58.6	46.4	51.5	47.7	47.7	52.2	43.9	52.2	42.4	48.5	35.8
	/z/	72.7	74.4	72.6	73.2	74.2	72.3	75.7	74.9	70.3	73.5	68.1	68.3
	/v/	69.3	69.5	69.5	71.3	69.2	68.2	66.6	67.7	68.1	63.7	58.2	58.5
Glottal fricative	/hh/	83.8	88.2	84.3	85.0	87.3	84.9	84.3	85.2	78.8	83.2	78.4	76.8
Unvoiced fricatives	/s/	81.8	81.9	82.8	81.9	81.2	82.7	80.1	77.5	82.1	78.4	71.7	76.2
	/th/	59.2	55.8	59.4	59.3	61.2	54.3	54.9	56.7	56.0	49.4	54.4	42.3
	/f/	89.2	88.0	88.1	84.1	84.1	87.2	86.4	85.3	84.6	85.7	80.2	79.8
	/sh/	87.2	85.8	87.1	86.3	87.8	86.6	87.1	88.6	89.0	88.1	82.9	83.6
Affricatives	/ch/	76.9	82.5	78.4	74.8	77.6	80.2	78.4	73.4	73.8	74.9	64.6	74.7
	/jh/	75.6	74.6	72.3	77.8	78.1	71.5	76.9	76.5	75.2	77.7	69.7	70.8
Average correct [%]		75.8	75.2	75.7	75.0	75.4	74.7	73.7	72.9	72.2	73.8	68.5	66.1

Table 5.5 American-English monophone recognition results for twelve speech parameterizations – percentage of correctly recognized monophones for the case of vowels, diphthongs, and semivowels

American English (39 monophones) vowels, diphthongs, semivowels		PLP-FB18	HFCC-FB40	MFCC-FB23	HFCC-FB28	MFCC-FB40	HFCC-FB23	WPF-FD	WPF-SBC	WPF-OBJ	LFCC-FB40	WPF-OVL	WPF-ACE
Front vowels	/iy/	**84.9**	84.3	**84.6**	84.5	82.4	84.4	82.7	78.7	82.5	83.1	74.6	83.1
	/ih/	53.5	**54.4**	**53.6**	53.4	52.3	52.3	51.8	45.5	48.0	48.3	45.2	47.3
	/ae/	75.4	**75.9**	75.6	75.0	73.9	74.4	74.6	69.1	73.5	72.1	63.7	75.4
	/eh/	**60.8**	60.7	58.2	57.9	59.1	58.4	55.6	48.4	55.7	56.2	47.0	56.8
	/er/	**78.0**	77.4	**78.0**	77.5	76.5	74.9	75.4	77.3	76.8	74.7	69.6	74.3
Back vowels	/uh/	**49.2**	46.6	41.8	43.0	41.0	43.5	42.4	35.8	41.9	41.1	32.4	42.4
	/ao/	70.3	71.3	72.0	72.1	71.7	72.4	70.8	68.7	**73.3**	71.5	64.8	**73.0**
	/ah/	**51.0**	49.3	49.6	49.4	49.0	**50.1**	42.3	46.2	43.4	42.4	39.4	44.1
	/aa/	67.2	**69.0**	**68.0**	66.3	67.9	65.7	65.9	58.9	64.6	63.2	61.0	67.7
	/uw/	68.2	69.4	68.2	68.6	69.2	**70.5**	66.0	64.4	64.4	68.4	57.5	**71.1**
Diphthongs	/ay/	83.1	83.1	**84.3**	**83.3**	82.8	82.3	82.7	74.2	82.6	82.9	73.2	78.7
	/oy/	**89.5**	88.4	88.3	85.4	**88.9**	88.6	85.4	78.6	83.0	83.1	82.0	86.3
	/ey/	81.3	79.2	81.2	81.8	80.5	80.5	81.1	72.9	**82.2**	80.6	72.3	**83.9**
	/ow/	70.9	71.1	**72.4**	69.5	70.6	**71.5**	64.0	62.0	68.4	63.8	57.6	68.6
	/aw/	69.0	**73.8**	70.4	67.5	**73.1**	70.1	68.8	61.3	66.8	66.2	58.4	68.8
Semivowels	/w/	84.5	85.8	85.4	**86.0**	85.3	84.4	85.3	84.7	**87.4**	85.6	80.9	85.1
	/l/	69.2	66.9	68.4	**69.6**	67.6	**68.8**	66.5	68.4	65.4	65.4	62.8	61.6
	/r/	74.8	**76.4**	74.0	**76.2**	75.7	76.1	71.3	74.8	71.7	70.6	60.1	72.4
	/y/	**81.8**	79.0	81.2	80.1	80.5	**82.6**	77.6	81.7	79.0	77.5	72.7	77.4
Average correct [%]		**71.7**	**71.7**	70.9	**71.3**	70.9	71.1	69.0	69.0	68.2	65.9	69.4	61.9

Again, the PLP-FB18 speech parameterization was observed to offer the highest average monophone classification rate and the HTK MFCC-FB23 showed the second best performance. The WPF-FD speech features were observed to offer some advantage over the MFCC-FB23 for the unvoiced stop /p/, the voiced fricatives /z/, /zh/, and also for the affricate /jh/, which partially supports the conclusions in Farooq and Datta (2001).

The WPF-SBC, although on average inferior to the PLP-FB18 and to the various MFCC and HFCC speech features, demonstrated the best classification accuracy for the unvoiced fricative /sh/ and the second-best classification rate for the voiced fricative /z/. Quite surprisingly, the WPF-OBJ demonstrated advantage over all other speech features for the voiced plosive /b/, voiced fricative /dh/, and the unvoiced fricative /sh/; the back vowel /ao/ and the semivowel /w/; and the second-best classification rate for the voiced fricative /zh/ and the diphthong /ey/.

Even more surprisingly, the WPF-ACE speech features, which show the lowest average classification rate, outperformed all other speech features on the back vowel /uw/ and the diphthong /ey/ and showed the second-best result for the back vowel /ao/. Some sufficient detailed explanation of the aforementioned results would require in-depth analysis of the properties of these phones and analysis of the related properties of each speech parameterization scheme.

Among the factors that influence the performance of speech parameterization and that might contribute to the observed ranking results and the obvious gap between the recognition accuracy obtained for the DFT- and DWPT-based speech features is the choice of speech pre-processing steps. For instance, one quite important factor could be the choice of speech frame size and the fact that the DFT-based methods make use of Hamming window in the speech pre-processing, which halves the effective frame size when compared to the rectangular window used in the DWPT-based methods. Shorter frame size facilitates the recognition of short phones and phones which have transient states, while larger frame size facilitates the recognition of long-duration phones with steady states.

Another factor is the influence of the properties of the basis functions: the DFT seems advantageous in the representation of relatively long steady periodic signals, such as the nasals, while DWPT might have advantage for some fricatives of phones with transients.

It is obvious from the results discussed above that the monophone classification accuracy could certainly be improved to some degree by the combined use[43] of few complimentary DFT- and DWPT-based speech parameterizations.

---

[43] This could be the "best-selection" method, or a fusion of the outputs of several parallel recognizers, or another collaborative method.

6 Evaluation on the Speech Recognition Task

Section 6 offers a direct comparison of the practical worth of six DFT- and five DWPT-based speech parameterizations on the task of continuous speech recognition. The DFT-based speech features of interest are the LFCC-FB40, MFCC-FB40, HFCC-FB23, HFCC-FB28, HFCC-FB40, and PLP-FB19, and the DWPT-based speech features are the WPF-SBC, WPF-FB, WPF-OVL and two versions of the WPF-OBJ. In all experiments, we made use of the well-known and widely used TIMIT database, which offers a standardized evaluation setup, and well-understood experimental protocol.

6.1 The Speech Recognition Task

Speech recognition is an intrinsic part of the spoken interaction process, regardless whether humans or machines are involved. When discussing about the automatic speech recognition by machines, we refer to only that part of the process which converts the captured speech signal to a sequence of written words, while the interpretation of the meaning of words is inseparable part of the speech recognition process when performed by humans.[44]

Depending on the way speech is pronounced, the automatic speech recognition process is referred to as *isolated word* recognition,[45] when the goal is the recognition of separately spoken commands or digits, or as *continuous speech* recognition, when the goal is to transcribe typical conversational human speech. Depending on the number of words that the speech recognition process can handle, we speak about *small vocabulary* (typically 10–100 words spoken as isolated words), *large vocabulary* (typically 1000–10000 words, spoken as continuous speech), or *very large vocabulary* (over 20000 words, spoken as continuous speech). Depending on the capability of a system to reliably recognize speech from one person or large number of people, we speak about *speaker-dependent* or *speaker-independent* speech recognition.

The development of automatic speech recognition technology with the use of digital computers has more than 50 years of history, and crediting all major contributors, whose creativeness facilitated the advance of speech recognition technology is beyond the reach of this book. However, there are a number of

---

[44] In order to research the human speech recognition process isolated from the interpretation part, researchers typically use syllables or combinations of phonemes which do not encode linguistic meaning in the language spoken by the test subjects engaged in the specific study.

[45] Isolated word recognition is used for simplifying the speech recognition task and improving the performance of speech recognition technology in certain practical applications, where such simplification is tolerated.

excellent books devoted to speech technology, and the interested reader shall refer to O'Shaughnessy (1987), Deller et al. (1993), Huang et al. (2001), etc., for further details[46] on the stochastic approach for speech recognition, which dominated the area in the past 30 years.

Two types of errors are often used as measure of the speech recognition performance in the continuous speech recognition task: *word recognition rate* (WRR) and *sentence recognition rate* (SRR), or their mirror representations in terms of error: *word error rate* (WER) and *sentence error rate* (SER), respectively. Although other less strict[47] measures are often used in practical applications, in this section we present the results in terms of WER and SER. Following the notations of the widely used open-source HMM ToolKit (HTK) (Young et al. 2006) we describe the WER as

$$\text{WER} = \frac{D_\text{w} + S_\text{w} + I_\text{w}}{N_\text{w}} \cdot 100\%, \tag{6.1}$$

where N_w is the total number of words in the test dataset, D_w is the number of *deletion errors* (words that were omitted in the obtained transcription when compared to the reference ground-truth transcription), S_w is the number of *substitution errors* (words that were misrecognized as another word), and I_w is the number of *insertion errors* (words that were inserted in the transcription produced by the speech decoder but do not exist in the reference ground-truth transcription of the spoken utterance). However, as the values of the WER are not bound in the range [0, 100] % and can take values larger than 100%, the WER is not an intuitive measure of accuracy. Thus, the percent of correctly recognized words, which ignores the insertion errors, is often provided as an additional measure. Finally, the SER is computed as the ratio between the number of sentences that contain at least one error on word level and the total number of sentences, multiplied by 100.

In the comparison of the speech recognition performance obtained for the various speech parameterizations evaluated in the current Sect. 6, we make use of both the WER and the SER as accuracy measures. Although they are not independent variables, when used together they provide the opportunity of mapping the results from each experiment on the SER – WER plane. This is considered as a more illustrative representation when compared to using only the WER ranking of the speech parameterization methods.

---

[46] A brief account of the difficulties related to the speech recognition problem is available in Deller et al. (1993), Section 10.1.

[47] In spoken dialogue interaction and command-and-control speech technology applications, another less strict measure, known as *task completed*, is often used. It does not account for the individual mistakes on word level: deletion, insertion, or substitution errors, but accounts only for the outcome of speech interpretation, that is, if the system got the meaning of the message correctly. This is somehow fair, as humans also are able to understand the meaning of a sentence or a phrase even when an unknown word is present or a word is omitted.

6.2   Experimental Protocol

A common experimental setup and protocol were used in all speech recognition experiments[48] reported in Sect. 6.3. The experimental setup is based on the well-known CMU Sphinx-3 speech recognizer (Lee et al. 1990), and the widely used TIMIT database (Garofolo 1998). This offers the opportunity for a direct comparison with other experimental work published on the same setup.

In brief, TIMIT consists of studio-quality speech recordings of phonetically balanced sentences from 630 American-English speakers representing eight dialects. For each speaker there are ten speech recordings, sampled at 16000 Hz with 16-bit precision per sample, which are divided into balanced training and testing sets of 462 and 168 speakers, respectively. The total number of words in TIMIT test subset is 14553, and the total number of test sentences is 1680. In the experiments, we took advantage of the TIMIT dictionary provided with database and a set of 38 phones {$aa, ae, ah, ahr, aw, ay, b, ch, d, dh, eh, er, ey, f, g, hh, ih, iy, jh, k, l, m, n, ng, ow, oy, p, r, s, sh, t, th, uh, uw, v, w, y, z$}.

The speech parameterization methods of interest here are the LFCC-FB40, MFCC-FB40, HFCC-FB23, HFCC-FB28, HFCC-FB40, and PLP-FB19, and the DWPT-based WPF-SBC, WPF-FB, WPF-OVL, and WPF-OBJ. In all experiments, only the first 13 cepstral coefficients were kept and the feature vector was obtained by appending the first and second time derivatives. This way every speech frame of 16 ms was represented by a 39-dimentional feature vector. The feature vectors were computed 100 times per second (a sliding window with a step of 10 ms). No automatic gain control or variance normalization was applied to the feature vectors. Further details are available in Mporas et al. (2007).

For each of the aforementioned speech parameterizations of interest, a separate Sphinx-3 acoustic model was trained. The acoustic models used three-state HMMs with a non-emitting terminating state. An HMM model was built for each of the 38 monophones plus one model for silence. Next, context-dependent untied triphone models were trained for every triphone that had occurred at least eight times in the training data. In total, 1000 senones were trained.[49] Finally, the context-dependent tied models were trained. Each HMM state is modeled by a mixture with eight Gaussian components. A tri-gram language model built from all TIMIT sentences is used. It was built with the help of the CMU Language Modeling Toolkit.[50] During the speech recognition phase, we made use of silence word probability of 1.0 and a language model weight of 9.5. The other settings of the decoder have their default values, as specified in the Sphinx-3 documentation.

---

[48] The comparative evaluation of the speech features presented in the current Sect. 6 is based on the work of Mporas et al. (2007).

[49] RobustGroup's Open Source Tutorial – Learning to use the CMU SPHINX Automatic Speech Recognition system, http://www. speech.cs.cmu.edu/sphinx/tutorial.html

[50] The CMU-Cambridge Statistical Language Modeling Toolkit, v2, http://www.speech.cs.cmu. edu/SLM/toolkit_documentation.html

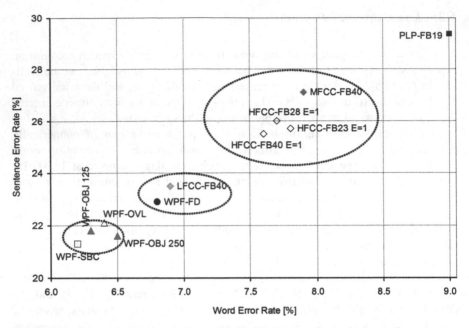

Fig. 6.1 Speech recognition accuracy on the TIMIT database for various speech parameterizations. The speech features whose performance is statistically the same are grouped with ellipses

6.3   Comparative Evaluation on the Speech Recognition Task

All speech feature sets of interest were processed[51] in a uniform manner as described in the Sect. 6.2. Here, the MFCC-FB40 are considered as the baseline speech parameters, as they are the default speech parameterization used in the CMU Sphinx-3 speech recognizer.

In Fig. 6.1, we present the experimental results for the DFT-based LFCC-FB40, MFCC-FB40, HFCC-FB23, HFCC-FB28, HFCC-FB40, and PLP-FB19, and the DWPT-based WPF-SBC, WPF-FB, WPF-OVL, and WPF-OBJ speech features on the SER-WER plane. Specifically, the speech recognition result for the LFCC-FB40 is shown with a "♦" (Orange color), the MFCC-FB40 with a "♦" (Red color), the HFCCs with a "◇" (no fill), the PLP-FB with a black square, the WPF-SBC with a framed light color square, the WPF-FD with black circle, the WPF-OVL with a triangle and no fill, and the two versions of the WPF-OBJ with filled triangles in dark color. The designation WPF-OBJ 250 stands for the WPF-OBJ as they were described in Sect. 4.4, and the designation WPF-OBJ 125 corresponds to a slightly modified frequency decomposition, where the frequency range [4000, 7000] Hz is covered by sub-bands with bandwidth of 125 Hz, instead of the 250 Hz used in the first version.

---

[51] The analysis in Sect. 6.3 is based on the results reported in Mporas et al. (2007).

As presented in Fig. 6.1, all speech parameterizations evaluated here outperformed the baseline MFCC-FB40, except of the PLP-FB19 cepstral coefficients, which demonstrate higher WER and SER values. Here, the advantageous accuracy of the HFCCs, WPF-FD and WPF-SBC was expected, as it is in agreement with the results reported by the corresponding authors (Skowronski and Harris 2004; Farooq and Datta 2001; Sarikaya et al. 1998) in different experimental setups. The WPF-SBC showed a relative reduction of the WER by more than 20% when compared to the baseline MFCC-FB40.

On the other hand, the good results obtained for the WPF-OVL and the WPF-OBJ 125 and WPF-OBJ 250 are quite unexpected as these features were initially designed for the needs of speaker recognition, and their frequency division was optimized in an objective manner on the speaker verification task. Another interesting observation is that the LFCC-FB40 speech parameterization, which uses a filter-bank of equal-bandwidth filters with linear spacing of their central frequencies, outperformed the HFCCs, PLP-FB19, and MFCC-FB40, which all possess a frequency warping inspired by the properties of the human auditory system.

This ranking of the speech parameterization methods contradicts the ranking obtained on the monophone recognition tasks. The last could be seen as puzzling and nonintuitive, as in the speech processing community the monophone recognition task was once considered as a first step of the speech recognition process. By that reason, the speech feature evaluation ranking once obtained on the monophone recognition tasks were often directly generalized as valid to the speech recognition task and by historical reasons to the speaker recognition tasks. As we will also see in Sect. 7 such a generalization does not hold.

In order to assess the statistical significance of obtained results, the t-test was performed for every pair of results (for details refer to Table 3 in Mporas et al. 2007). In Fig. 6.1, the speech recognition results within each dashed ellipse are not statistically different from those of the other members of that group. Therefore, as shown in the figure, one can say that the difference between the speech recognition results for the WPF-SBC, WPF-OBJ, and WPF-OVL is not statistically significant. The same holds for the group WPF-FD and LFCC-FB40, and for the group of HFCCs and MFCC-FB40. However, the results from the t-test demonstrate that the difference in the speech recognition accuracy obtained for these three groups is statistically significant.

In conclusion, based on these experimental results, one can generalize that the DWPT-based speech parameterizations studied here outperform the baseline MFCC-FB40 and the other DFT-based speech parameterizations. The superior speech recognition accuracy obtained for the DWPT-based speech features is due to: (i) the balanced time–frequency resolution these wavelet packet trees provide when compared to the uniform frequency resolution of the DFT-based ones and (ii) to the more suitable (for analysis of nonstationary speech signals) basis functions, which are more reasonable choice when compared to the cosine functions.

7   Evaluation on the Speaker Verification Task

In this section, we investigate the appropriateness of various speech parameterizations and 60 different subsets of speech features on the speaker verification task.
First, we evaluate six DFT-based speech features (MFCC-FB20, HTK MFCC-
FB24, MFCC-FB32, HFCC-FB19, HFCC-FB24, and HFCC-FB29) and afterward
compare the best of them with four DWPT-based speech parameterizations (WPF-
SBC, WPF-FD, WPF-OVL, and WPF-OBJ). We show that some specific subsets of
the DWPT-based speech features outperform the commonly used MFCC. Here, we
make use of the experimental protocol developed for the needs of the annual
NIST[52] speaker recognition evaluation (SRE) campaigns (NIST 2001).

7.1   The Speaker Verification Task

The speaker verification process, based on an identity claim and a sample of
speaker's voice, provides an answer to the unambiguous question: *"Is the present
speaker the one s/he claims to be, or not?"* The output of the verification process is a
binary decision "Yes, s/he is!" or "No, s/he is not!". The actual decision depends on
the degree of similarity between the speech sample and a predefined model for the
enrolled user, whose identity the speaker claims. When an enrolled user claims her/
his true identity, we designate the input utterance as a *target* trial. When a nonuser
addresses a speaker verification system, or when an enrolled user claims identity
belonging to another user, we denote that utterance as a *nontarget* trial. The
nontarget trials are also referred to as *impostor* trials.

Thus, in the speaker verification problem,[53] we consider two hypotheses – either
the input speech originates from the same person, whose identity the speaker
claims, or it originates from another person, who has different identity. In order
to test each of these two hypotheses, we build an individual expert, that is, a model
for each enrolled user. In fact, each expert incorporates not one but two models: one
built from the voice of the enrolled user, and another one representing the rest of the
world. The latter one is also designated as a *reference* model or *background* model.
Since the reference model has to be sufficiently general, it is built from the speech
data of a large number of speakers.

With respect to linguistic contents of speech, the speaker recognition process can
be *text-dependent* or *text-independent*. The text-dependent speaker verification
systems examine the manner in which a specific password or a system-prompted

---

[52] The abbreviation NIST stands for the National Institute of Standards and Technology of the USA.

[53] A nice introduction to the speaker recognition problem is available in Campbell (1997), and a
recent overview of the state-of-art speaker verification technology is offered in Kinnunen and
Li (2010).

sequence is pronounced. In the text-independent scenario, the talker is not restricted in any way, and as soon as the identity claim is provided, s/he is free to speak naturally, without any vocabulary restrictions and often without being cooperative. Here, we consider the text-independent scenario.

Two types of errors can occur in the speaker verification process. The first one called a *false rejection* (FR) error occurs when the true target speaker is falsely rejected as being an impostor, and as a result, the system *misses* to recognize an attempt belonging to the true authorized user. The second type called a *false acceptance* (FA) error occurs when a tryout from an impostor is accepted as if it comes from the true authorized user. The latter error is also known as a *false alarm*, because a nontarget trial is accepted as a target one. The FR and FA are often employed together to characterize the speaker verification accuracy. In the following subsections, we make use of the cost-based performance measure, C_{Det}, defined as in (7.1), for assessing the speaker verification accuracy. It is defined (NIST 2001) as a weighted sum of the false acceptance and false rejection error probabilities, designated as $P(FalseAlarm|NonTarget)$ and $P(Miss|Target)$, respectively:

$$C_{Det} = C_{Miss}P(Miss|Target)P(Target) + C_{FalseAlarm}P(FalseAlarm|NonTarget)$$
$$(1 - P(Target)), \tag{7.1}$$

where the parameters C_{Miss} and $C_{FalseAlarm}$ are the relative costs of detection errors, and $P(Target)$ is the *a priori* probability of the specified target speaker.

The cost measure C_{Det} is further normalized as:

$$C_{Norm} = C_{Det}/C_{Default}, \tag{7.2}$$

where

$$C_{Default} = \min\{C_{Miss}P(Target), C_{FalseAlarm}P(NonTarget)\} \tag{7.3}$$

for making its values more intuitive. Here, $C_{Default}$ represents the zero value (a system providing no information), which is the cost obtained without processing the data, always making the same decision – either accept or reject. Finally, the range of values received by C_{Norm} is between zero, for a system that makes no mistakes, and a positive constant that depends on the ratio of the products $C_{Miss}P(Target)$ and $C_{FalseAlarm}P(NonTarget)$, for a worthless system. The *actual decision cost, DCFact*, is the decision cost C_{Norm} computed after the final decision is made. Thus, the *DCFact* depends not only on the quality of modeling, but also on the relevance of a priori estimated speaker-independent threshold. On the other hand, the *optimal decision cost, DCFopt*, gives an impression about the maximum potentially achievable performance of a system for the specific models, when the optimal speaker-independent threshold is applied.

The accuracy of the speaker verification systems is often evaluated at the *equal error rate* (EER) point, which is a more intuitive measure of performance. The EER is computed as the mean between the FA and FR near the point, where the false rejection and the false acceptance error probabilities are equal. The EER assumes equal weights for the speaker verification cost model parameters $C_{Miss} = C_{FalseAlarm} = 1$, as this is a more intuitive setup. In practice, these costs are application-dependent. Their ratio could vary from one application to another in the range of 1:10–10:1, depending on whether the emphasis is placed on the security of access or on the comfort of use. It has to be emphasized that the EER point decision offers a more balanced, but also a too optimistic estimation of the speaker verification accuracy.

In the current Sect. 7, we make use of both the *DCFopt* and the EER as measures of the speaker verification accuracy. Used together they provide the opportunity of mapping the results from each experiment on the *DCFopt* – EER plane, which offers a distinctive representation of every experimental result.

7.2   Experimental Protocol

A common experimental protocol was followed in all validation experiments according to the rules described in the 2001 NIST SRE plan (NIST 2001). In brief, the male part of the NIST 2001 SRE one-speaker detection cellular-network database was used. About 40 s of voiced speech on average were detected in the training portion of the database, which contains a single 2-min recording for each of the 74 male speakers. The speech features computed for the voiced speech frames were next used for training the clients' models. The common reference model was created by exploiting the male training speech available in the NIST 2002 SRE database (NIST 2002). Approximately, 1 h and 40 min of voiced speech was available for that purpose.

After training, the user models were tested carrying out all male speech trials as defined in the complete one-speaker detection task (the index file "detect1.ndx" provided with the database). Each speaker verification experiment included 850 target and 8500 impostor trials with duration from 0 to 60 s of speech, and employed recordings from all transmission channel types. A comprehensive description of the evaluation database and evaluation rules is available in the 2001 NIST SRE plan (NIST 2001).

In order to accommodate the various speech parameterization schemes to sampling rate of 8000 Hz, which the use of the NIST SRE databases imposes, we excluded from all filter-banks these filters which spread beyond the 4000 Hz boundary. Thus, in the experiments with the MFCC-FB20 (Davis and Mermelstein 1980), we used 19 filters – 10 with linearly spaced center frequencies and 9 with logarithmically spaced ones. Next, following the instructions in Young (1996), we used a filter-bank of 20 filters for computing the HTK MFCC-FB24 features. In the experiment with Slaney's MFCC-FB40, we kept the first 32 filters, which cover the frequency range [133, 3954] Hz, and we refer to this version as to MFCC-FB32. In the experiment with the HFCC-E-FB29 (using E-factor $E = 1$) we tested various number of filters (19, 24, 29) to cover the frequency range of [133, 4000] Hz. In the case of the DWPT-based speech features (WPF-FD, WPF-SBC, WPF-OBJ, and

WPF-OVL), we made use of the narrowband version [125, 4000] Hz of each speech parameterization as these are described in Sect. 4. Here, the DWPT decomposition was performed with a common wavelet function, Battle-Lemarié of order 5, for all WPF. Therefore, the various WPF differ solely in the wavelet packet tree and subsequently in the frequency warping along the frequency range [125, 4000] Hz. Dynamic range normalization was used for all speech features discussed in Sect. 7.3 but is not applied to the DWPT-based speech features evaluated in Sect. 7.4.

All DFT-based and DWPT-based speech features were evaluated on the 2001 NIST SRE database by means of the PNN-based text-independent speaker verification system[54] described in Ganchev et al. (2002a, b). It offers fast training times, which facilitated the evaluation of multiple speech parameterizations and the experiments with multiple subsets of speech features.

7.3 Comparison Among Six MFCC Implementations

The MFCC implementations MFCC-FB20, HTK MFCC-FB24, MFCC-FB32, HFCC-FB19, HFCC-FB24, and HFCC-FB29, outlined in Sects. 3.2–3.5 were evaluated in the common experimental setup described in Sect. 7.2.

Figure 7.1 shows the experimental results for these MFCC implementations. The left-hand bottom corner in the figure corresponds to the lowest $DCFopt$ and EER, and thus to the best speaker verification accuracy. In contrast, the right-hand upper corner in the figure corresponds to the worst speaker recognition results, as the $DCFopt$ and EER have their highest values there. On the figure, each experiment is represented with a single mark, and has a distinctive marker and a label with the name of the speech parameterization method and the subset of speech features used. As the figure shows, there is no significant difference among the results for the HTK MFCC-FB24, Slaney's MFCC-FB32, and the HFCC-E-FB29 $E = 1$ (with 29 filters covering the range [0, 4000] Hz), which was expected based on previous experience. Next, the MFCC-FB20 computed as in Davis and Mermelstein performed slightly worse, and finally, the highest EER was observed for the HFCC-E-FB24 and HFCC-E-FB19 with $E = 1$. In fact, Skowronski and Harris (2004), suggested 29 filters for the frequency range [0, 6250] Hz, which corresponds to a filter-bank of 24 filters for

---

[54] The accuracy of this speaker verification system is inferior when compared to the present state-of-art systems as the JFA GMM-UBM (Kenny et al. 2008) or GSV-SVM (Campbell et al. 2006), especially in mismatched train-test conditions. However, since here we aim at comparing the practical usefulness of various feature extraction techniques rather than optimizing an absolute speaker verification accuracy, we take advantage of the fast training times of the PNN-based system which are of three orders of magnitude faster, when compared to these of the aforementioned systems. Upon the availability of sufficient computational resources, the interested reader may wish to take advantage of the open-source ALIZE speaker recognition toolkit (Fauve et al. 2007) or another implementation of the aforementioned state-of-art speaker verification systems and evaluate the performance of certain speech parameterizations.

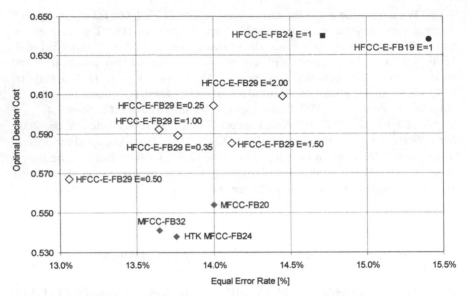

Fig. 7.1 Speaker verification results in terms of EER and DCFopt for various DFT-based speech features

the frequency range [0, 4000] Hz but we also experimented with 29 filter in the frequency range [0, 4000] Hz. Interestingly, the speaker verification experiments demonstrated that 29 filters for the frequency range [0, 4000] Hz offers a lower EER than the versions with 24 or 19 filters. We deem the reason for this result is (at least in part) in the insufficient overlapping between the first few filters in the HFCC-E filter-bank, especially when the number of filters is small. Small number of filters results in bad frequency resolution at low frequencies due to the lack of overlapping among the neighboring filters (or due to their insufficient overlapping), which leaves some sub-bands underrepresented in the feature vector. In addition, examining the results for MFCC-FB40 and HFCC-E-FB29, we see that the larger number of filters in the filter-bank facilitate the achievement of a better speaker verification accuracy, as the smoothing of the speaker-specific details in the spectrum decreases. The only exception here is the result of HTK MFCC-FB24 – apparently other factors influence the speaker verification accuracy as well.

In order to investigate the importance of the E-factor and its impact on the accuracy obtained with the HFCC, we experimented with various values of $E = \{0.25, 0.35, 0.5, 1.0, 1.5, 2.0\}$. In the initial experiments with a filter-bank of 24 filters for the frequency range [0, 4000] Hz, it was observed that $E = 1$ provides the lowest EER.

However, the most appropriate value for the E-factor was found to depend on the design of the filter-bank. In order to illustrate this, in Fig. 7.1, we present results for the best HFCC-E-FB29 – with 29 filters in the frequency range [0, 4000] Hz and various values of the E-factor. As the figure shows, for this specific filter-bank the

E-factor $E = 0.5$ provides the lowest EER. Any deviation from the value $E = 0.5$ in either direction was observed to increase the EER. We deem the reason for such behavior is that for lower values of the E-factor, the filters with the lowest center frequencies barely overlap, and thus the filter-bank resolution for these frequencies is low – threshold phenomena were observed. For higher values of the E-factor, the filters are very broad and thus they smooth far too much the speaker-specific details in the spectrum, which are important for distinguishing among different speakers. The last was reported useful for the robustness of speech recognition (Skowronski and Harris 2004) but does not favor the speaker recognition task. Among all speech features evaluated here the MFCC-FB40 and the HTK MFCC FB-24 were found to offer the lowest decision cost, $DCFopt$.

In conclusion, we can say that, as expected, the speaker verification accuracy did not vary vastly when different approximations of the nonlinear pitch perception of human are used in the MFCC computation. Exceptions are only the HFCC with filter-bank of 19 and 24 filters, which offer much inferior accuracy. Furthermore, the experimental results support the hypothesis that regardless of the specific filter-bank design, the use of a larger number of filters favors the speaker verification accuracy. Besides the number of filters in the filter-bank, the overlapping among the neighboring filters also proved to be a sensitive parameter. Increase or decrease of this overlapping beyond a given range increases the EER.

7.4 Comparison Among Four DWPT-Based Speech Features

In a thorough evaluation on the speaker verification task, four DWPT-based speech parameterizations (WPF-FD, WPF-SBC, WPF-OBJ, and WPF-OVL), outlined in Sect. 4, were compared with the baseline MFCC-FB32.

In Fig. 7.2 we show the speaker verification results for 60 different experiments. These experiments differ either in the type of speech features or in the subset or cepstral coefficients that was employed. For the experiments with the MFCC and HFCC, we also show the results with post-processing of the cepstral coefficients, as it was observed that the cepstral mean subtraction (CMS) and the dynamic range normalization (DRN), that is, variance normalization, contribute significantly to improving the speaker verification accuracy. In Fig. 7.2, we use different shapes of the markers as follows: "●" for the WPF-FD, "▲" for the WPF-OBJ and WPF-OVL, "■" for the WPF-SBC, "◆" for the MFCC, and "◇" for the HFCC speech parameterizations.

As Fig. 7.2 shows, the best subset among all speech features is the WPF-OBJ {4:40}, which stands for the WPF-OBJ speech parameterization computed for $R = 41$ but with the first three cepstral coefficients discarded. The speech features, WPF-OBJ {4:40} demonstrate the lowest $DCFopt$ among all subsets, and together with the subsets WPF-OBJ {4:35} and WPF-OBJ {4:50} demonstrate the best speaker verification accuracy in terms of both EER and $DCFopt$. Some of the other speech feature subsets, such as the WPF-OBJ {4:30}, WPF-OVL {4:35},

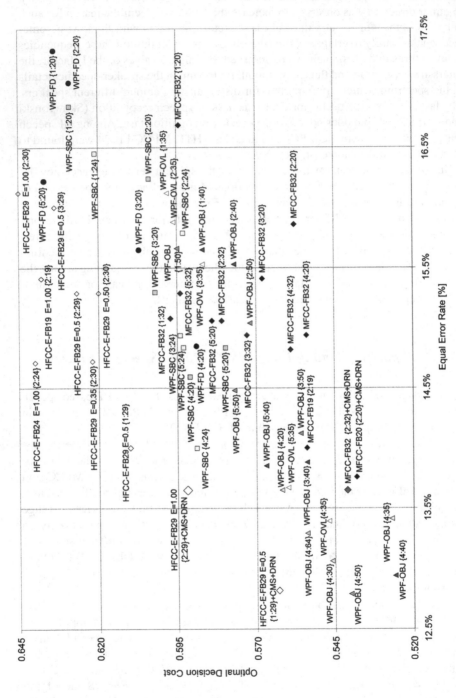

Fig. 7.2 Results for the WCL-1 system on the NIST2001 SRE one-speaker detection task: male trials, voiced speech only, frame size 32 ms, skip 16 ms

and WPF-OBJ {4:64}, also offer a balanced speaker verification accuracy in terms of both the EER and the *DCFopt*.

We deem that the gain of speaker verification accuracy, which is obtained after removing the first few coefficients from the feature vector, is due to:

- Their sensitivity to the linguistic content of the speech segment
- The high sensitivity of these coefficients to mismatch between training and testing conditions, that is, due to different speech transmission channels, microphones, recording setup, etc.
- Their relatively bigger dynamic range and variance when compared to the coefficients with larger indexes

The last holds to some extent for the DFT-based features but not that much for the DWPT-based speech features, where CMS and DRN were not confirmed to contribute to improving of the speaker verification accuracy (Ganchev 2005).

As it is well known, the value of the first cepstral coefficient (the one with index zero) is proportional to the logarithm of the energy of the corresponding speech frame, and therefore, it is much dependent on the sound acquisition setup: the distance between the mouth and the microphone, the type of the transducer, on the speaking style and conditions, etc. Thus, discarding the first cepstral coefficient from the feature vector is a common practice for reducing the dependence on the speech acquisition setup.

Furthermore, the value of the second cepstral coefficient is proportional to the balance between the lower- and upper-frequency halves of the speech spectrum. Thus, the values of the second cepstral coefficient depend much on the phonetic content of the speech segment, which is not advantageous in the text-independent speaker verification task that we consider here. To some extent, the same comments hold for the third cepstral coefficient, which represents the balance between the first-second and third-fourth quarters of the spectrum, but here each quarter is a sub-band of 1000 Hz (for sampling rate of 8000 Hz). In this way, the value of the third cepstral coefficient is influenced to some extent by the energy distribution among the formants, which is characteristic for the different phones. Thus, the third coefficient is also affected (but to a lesser extent) by the phonetic content in the current speech frame. For the cepstral coefficients with index four and higher, these sub-bands become more narrow and finer and the speaker-characterizing portion of the information becomes more important than the phonetic-content depending parts.

Next, looking at Fig. 7.2, we see that the best DFT-based speech features, the MFCC-FB32 {4:32}, obtain EER that is higher than the one for the WPF-OBJ {4:40} with about 15%, in terms of relative difference. After applying a post-processing on the MFCC-FB32 features, the best result among the DFT-based speech features is obtained for the MFCC-FB32 {2:32} + CMS + DRN. However, even after the post-processing, the accuracy of the best speech feature set, MFCC-FB32 {2:32} + CMS + DRN, is still inferior when compared to the one obtained for the best WPF-OBJ subset, the WPF-OBJ {4:40}. Likewise, the CMS and DRN help for significant decrease of the EER and the DCFopt for the HFCC-E-FB29 feature subset but its accuracy is still much worse than the one of the best WPF-OBJ subsets. This advantage of the WPF-OBJ and the WPF-OVL can be explained with the fact that their filter-banks have higher frequency resolution and frequency division, which is optimized (within a certain range) in an objective manner for the task of speaker verification. These filter-banks lead to a higher resolution representation of the speech cepstrum, which preserves more speaker-specific details, when compared to the 20 to 32 filters used in the other speech parameterizations.

On the other hand, the worst speaker verification accuracy among all speech parameterizations discussed here, that is, the highest EER and *DCFopt* is observed for various subsets of the WPF-FD, HFCC, and WPF-SBC speech parameterizations. In the case of the WPF-FD speech features, which were purposely designed for the needs of speech recognition, we consider the small number of filters in the filter-bank as the main reason for their inferior accuracy on the speaker verification task. Specifically, the small number of filters means that the same frequency range is covered with wider filters, and thus the smoothing of speaker-specific spectral details is more significant.

The small number of filters also bounds the number of unique cepstral coefficients that can be computed. As the experimental results show, in the general case, more coefficients contribute to a better speaker recognition accuracy (given sufficient amount of training data). To some extent, these comments also hold for the WPF-SBC speech parameterization. In the case of the HFCC, we deem that the inferior speaker verification accuracy is due to the suboptimal width and spacing of the filters in the filter-bank. Specifically, on one side, the lower-frequency filters barely overlap and thus important information is not captured, and on the other hand, the filters with higher central frequency become too wide, and thus they smooth too much of the speaker-relevant information. The adjustable E-factor controls the width of the filters but its lower bound value is restricted by the fact that narrow filters leave the lowest frequencies of the spectrum poorly covered.

In order to evaluate the potential benefits of the larger number of coefficients that the DWPT-based speech features offer, we show the speaker verification results for step-wise increasing of the dimension of the feature vector. For this purpose, we selected the results for the WPF-OBJ, as they demonstrated the best accuracy among all speech features considered here. Here, we do not aim at identifying the best accuracy that can be achieved on the NIST 2001 SRE database with a subset of the WPF-OBJ speech features, but instead to study the general trend.

Figure 7.3a, b show the EER and *DCFopt* for subsets of WPF-OBJ feature vectors that consist of the first 30, 35, 40, and 50 coefficients, as well as the entire set of 64 coefficients. Results for different subsets of cepstral coefficients, excluding the first, the first two, the first three, etc. are also shown. As the figure shows, the subset WPF-OBJ{4:40} has a significant advantage over the other subsets in terms of both EER and *DCFopt*. The other feature sets express either higher EER or *DCFopt*. The variable ranking in terms of EER and *DCFopt* for some subsets is due to the change in the slope of the DET performance plots (details available in Ganchev (2005), Sect. 4, Fig. 4.9). In Figs. 7.2 and 7.3 the effect from the change of slope in the DET plots is obvious. The same holds for the MFCC-FB32{4:20}

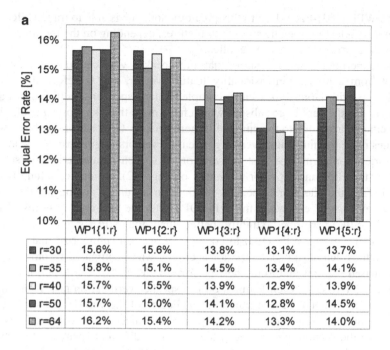

	WP1{1:r}	WP1{2:r}	WP1{3:r}	WP1{4:r}	WP1{5:r}
■ r=30	15.6%	15.6%	13.8%	13.1%	13.7%
□ r=35	15.8%	15.1%	14.5%	13.4%	14.1%
□ r=40	15.7%	15.5%	13.9%	12.9%	13.9%
■ r=50	15.7%	15.0%	14.1%	12.8%	14.5%
□ r=64	16.2%	15.4%	14.2%	13.3%	14.0%

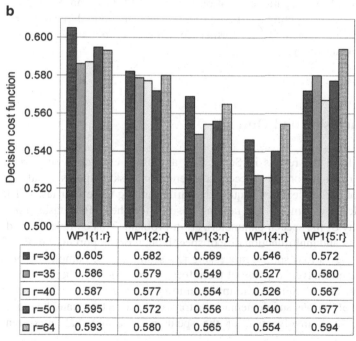

	WP1{1:r}	WP1{2:r}	WP1{3:r}	WP1{4:r}	WP1{5:r}
■ r=30	0.605	0.582	0.569	0.546	0.572
□ r=35	0.586	0.579	0.549	0.527	0.580
□ r=40	0.587	0.577	0.554	0.526	0.567
■ r=50	0.595	0.572	0.556	0.540	0.577
□ r=64	0.593	0.580	0.565	0.554	0.594

Fig. 7.3 EER and DCFopt for various subsets of the WPF-OBJ

and the WPF-OBJ{4:20}. Due to this phenomenon, one is able to trade EER versus *DCFopt* by selecting specific subsets of features, depending on the requirements of the specific speaker verification application.

The experimental results show that the speaker verification accuracy could benefit from a higher dimensionality feature vector, and therefore the cepstral coefficients with larger index carry complementary speaker-specific information. However, for the very large subsets, which have more than 40 cepstral coefficients, the accuracy drops due to the curse of dimensionality. On a larger database, which provides more training data, even larger feature vectors might turn out beneficial.

In brief, comparing the best subsets for all speech parameterization schemes (refer to the EER and *DCFopt* presented on Fig. 7.2), we can conclude that the WPF-FD exhibit the worst speaker verification accuracy among all the speech features evaluated here, while the WPF-OBJ offer the best one. In terms of EER, the WPF-SBC performed slightly better than the MFCCs but was entirely outperformed by the WPF-OBJ. In conclusion, we can summarize that the WPF-OBJ features demonstrated a superior accuracy when compared to other wavelet packet-based features and to the best-performing MFCC.[55] The superior accuracy of the DWPT-based speech features on the speaker verification task is deemed to the reason that the (i) wavelet function, (ii) design of the wavelet packet tree, and (iii) selection of frequency resolution were optimized in a systematic way to emphasize the dissimilarity between the voices of different speakers. Finally, the DWPT-based speech parameterization schemes WPF-OBJ and WPF-OVL offer the opportunity of computing a larger number of relevant nonredundant parameters for each speech frame, and this further contributes to obtaining a better speaker verification accuracy.

8   Conclusion and Outlook

As seen from the results in Sects. 5–7 and from evaluation of various speech parameterizations on the speech segmentation task (Mporas et al. 2008), the ranking of speech features is quite dissimilar among these four tasks. The significant differences among the ranking results for these four tasks illustrate the complexity of the speech parameterization problem. This complexity is due to the complexity and multifunctionality of speech but more importantly to the dissimilar needs of the various speech processing tasks. In fact, the complexity of speech is due to the multilayer structure of the information carried by speech, but also to the physiological and cultural variability among humans as source of speech, to the influence of various supplementary factors as the affective state, health condition, cognitive load, involvement of the speaker, environmental interferences, etc.

---

[55] Similar ranking was observed also on the Polycost speaker recognition database.

However, the complexity of speech is a less significant factor, and even when the variability due to this complexity is properly modeled and suppressed to a sufficient degree, the main difficulty originating from the dissimilar requirements to the speech parameterization process in the various speech processing tasks remains a major obstacle.

In this relation, one can figure out that the differences in the practical usefulness of the various speech parameterizations, depending on the task, operational setup, etc., are due to the fact that they rely on a rigid design and the filter-bank is with predefined resolution and sensitivity. The human audition seems to dynamically adapt the sensitivity and frequency resolution, depending on whether the interest is on the linguistic information in speech, on who speaks and what s/he feels, or whether the person listens to speech, music, or to the sounds of Nature. Such hypothesis could find some support in the joint interpretation of the research findings reported in Fletcher (1940), Zwicker (1961), Flanagan (1972), Chistovich (1985), and Hermansky (2003).

At present, the use of common speech parameterization process for the needs of nearly all speech processing-related tasks is an ordinary practice. This is supported by the common sense understanding that regardless of the task, humans use the same auditory apparatus for transforming speech to a suitable representation, which is then processed in the brain. However, although the functioning of the human auditory system is not as puzzling to us as it was to researchers in the beginning of the twentieth century, we still seem to lack in good understanding and in a compelling sensory-motor theory on how the human brain interacts with the auditory system during the different human activities, related to the various speech processing tasks. Thus, the development of speech processing technology which dynamically adapts to a particular task, environmental conditions, speaker state, etc. remains a remote goal.

During the past century of research on the human auditory apparatus, we gained an insight into the mechanisms of audition. In the same way, we may need another considerable effort in order to reach a better understanding on the manner in which brain interacts with the auditory apparatus, in order to achieve focused attention, adaptation to unseen conditions, and other important functionality which is inherent for the human audition. It will be of significant practical value to obtain a more comprehensive knowledge about the way in which information propagates from the auditory apparatus to the top layers in the brain, which realize the concept interpretation and understanding, and the mechanisms of information generalization. Such efforts may bring new ideas for biologically inspired speech processing and further advances of the speech parameterization process.

Nowadays, the DCT decorrelation of the filter-bank outputs is a common step in nearly all contemporary short-time spectrum-based speech parameterization methods, as it made the cepstral coefficients appropriate for use with the modern statistical speech processing techniques. Paradoxically, the DCT decorrelation of the filter-bank output is the least biologically plausible step of the speech parameterization process. Thus, although convenient for the statistical machine learning techniques used today, the DCT decorrelation might be excluded from tomorrow's speech parameterization methods, when the speech modeling paradigm is shifted toward more biologically plausible speech processing methods. These methods shall find the way to benefit from the inherent redundancy of information transmitted by the neighboring areas in the human cochlea and the long-term (up to few seconds) integration of this information in the brain. The emergence of biologically plausible speech and audio processing methods will most likely facilitate the efforts for achieving satisfactory noise robustness in a wider range of acoustic conditions but also will constitute a step toward obtaining a human-like perception and interpretation of sounds.

9 Links to Code and Further Sources of Information

Nowadays, *Wikipedia* is a fast growing and continuously updated repository of introductory materials, which offer a good starting point for obtaining a brief idea about a certain topic.[56] For example, the Wikipedia section devoted to the computation of the Mel frequency cepstral coefficients[57] got updated several times in the past few months while this book was in preparation. Thus, students and users of speech parameterization techniques could benefit from periodic visits to the Wikipedia relevant sections as, if certain information is not there today it might appear soon.

Both researchers active in the area and practitioners who make use of speech parameterization techniques could benefit from the interesting discussions in the online *Auditory list archives* (Research in Auditory Perception) hosted at the McGill University Web site.[58] There, one has the chance to find answers of important questions and to benefit from the guidance of the brightest researchers in the areas of speech processing and human audition. The author admires the Auditory list archives as a source of wisdom and knowledge and strongly suggests to the reader to search the online archives for answers of his/her questions before posting queries to this community. Another useful forum that can provide answers to the practical questions of students and practitioners is the *comp.speech.research* newsgroup. In addition, answers to frequently asked questions (FAQs) are posted there on periodic basis.

To reach beyond the coverage of this book, the interested reader shall search for related work on speech parameterization in the (online) archive repositories of the various speech processing-oriented scientific journals. Most often, new

---

[56] If the present speed of updating the Wikipedia content is preserved, just within few years Wikipedia might become a comprehensive and much demanded source of knowledge for students and practitioners who make use of speech parameterization techniques.

[57] At present, Wikipedia hosts a section devoted to the Mel frequency cepstral coefficients at URL: http://en.wikipedia.org/wiki/Mel-frequency_cepstrum

[58] Auditory list archives are presently hosted at URL: http://lists.mcgill.ca/archives/auditory.html

developments and advances in the area of speech parameterization appear in scientific journals published by Springer,[59] Elsevier,[60] IEEE,[61,62] and the Acoustical Society of America.[63] For up-to-date information and the latest advances in the area of speech parameterization, the reader might also benefit from attending speech processing-oriented conferences such as the annual ISCA-organized INTERSPEECH conference,[64] which at present is the biggest conference devoted to speech processing; the annual IEEE ICASSP,[65] which is the biggest conference devoted to signal processing; and the various other speech processing-related conferences organized across the globe (for up-to-date information and call for papers refer to the links in footnotes 65 and 66 at the bottom of this page and to the specialized Web sites[66,67]).

Source code and ready-to-use tools for most of the speech parameterizations described in this book is freely available on the Internet. The few speech parameterizations that at present are not covered by open-source code repositories can be easily derived from existing implementations of the Mel frequency cepstral coefficients, following the detailed description offered in Sects. 3 and 4.

At present, the MFCC speech features are widely used and their implementations are available as ETSI standards[68,69] and in any open-source or commercially available software that deals with speech processing. Among the well-supported open-source tools that compute MFCC are Praat,[70] SPro,[71] SFS,[72]

---

[59] International Journal of Speech Technology, http://springerlink.com/content/100275/

[60] Speech Communication, http://www.elsevier.com/wps/find/journaldescription.cws_home/505597/description#description

[61] IEEE Signal Processing Letters, http://ieeexplore.ieee.org/xpl/RecentIssue.jsp?punumber=97

[62] IEEE Transactions on Audio, Speech and Language Processing, http://ieeexplore.ieee.org/xpl/RecentIssue.jsp?punumber=10376

[63] The Journal of the Acoustical Society of America, http://scitation.aip.org/JASA

[64] Section Events at the International Speech Communication Association (ISCA) Web site, http://www.isca-speech.org/iscaweb/

[65] IEEE Signal Processing Society Web site, http://www.signalprocessingsociety.org/conferences/upcoming-conferences/

[66] WikiCFP: http://www.wikicfp.com/cpf/

[67] Wikipedia CFP: http://en.wikipedia.org/wiki/Call_for_papers

[68] ETSI ES 201 108, V1.1.2 (2000-4). ETSI Standard: Speech Processing, Transmission and Quality Aspects (STQ); Distributed Speech Recognition; Extended Advanced Front-end Feature Extraction Algorithm; Compression Algorithms; Back-end Speech Reconstruction Algorithm, April 2000, Chapter 4, pp. 8–11.

[69] ETSI ES 202 050, V1.1.5 (2007-1). ETSI Standard: Speech Processing, Transmission and Quality Aspects (STQ); Distributed Speech Recognition; Extended Advanced Front-end Feature Extraction Algorithm; Compression Algorithms; Back-end Speech Reconstruction Algorithm; January 2007, Section 5.3, pp. 21–24.

[70] Praat: http://www.fon.hum.uva.nl/praat/

[71] SPro: http://www.irisa.fr/metiss/guig/spro/

[72] Speech Filing System (SFS): http://www.phon.ucl.ac.uk/resource/sfs/

openSMILE,[73] etc. and the VoiceBox,[74] Auditory Toolbox,[75] Auditory/Cochlea Toolbox[76] for MATLAB.[77] In addition, there is a fast-growing repository of user-contributed source code at the MATLAB Central File Exchange[78] site, which the reader can check periodically in order to avoid duplication of efforts. Various MFCC implementations are also available in the open-source speech recognition platforms, such as the Cambridge HMM Toolkit (HTK),[79] Julius,[80] CMU Sphinx,[81] etc. Usually, these offer also implementation of the PLP cepstral coefficients as an alternative speech parameterization technique. The LFCC are computed by Praat and SPro. The HFCC code[82] is also publically available.

The author is not aware of any publically available ready-to-use source code that computes the DWPT-based speech parameterizations described in Sect. 4. However, their implementation is quite straightforward[83] as all functions needed are already available in the MATLAB Wavelet Toolbox[84] and the WaveLab ToolBox[85] for MATLAB, and thus, the reader can easily assemble the speech parameterization code by following the sequence of processing steps discussed in Sect. 4.

---

[73] openSMILE: http://opensmile.sourceforge.net/

[74] VOICEBOX: http://www.ee.ic.ac.uk/hp/staff/dmb/voicebox/voicebox.html

[75] Auditory Toolbox: http://cobweb.ecn.purdue.edu/~malcolm/interval/1998-010/

[76] Auditory/Cochlea ToolBox: http://www.it.uc.pt/~fp/func.html

[77] MATLAB is a trademark of The Mathworks Inc., http://www.mathworks.com/

[78] MATLAB Central File Exchange: http://www.mathworks.com/matlabcentral/fileexchange/

[79] HTK: http://htk.eng.cam.ac.uk/

[80] Julius: http://julius.sourceforge.jp/en_index.php

[81] CMU Sphinx: http://cmusphinx.sourceforge.net/

[82] HFCC code: http://www.cnel.ufl.edu/~markskow/

[83] The author's version of the DWPF-based speech parameterizations described in Sect. 4 is available at http://www.wcl.ece.upatras.gr/tganchev/

[84] Wavelet Toolbox: http://www.mathworks.com/products/wavelet/

[85] WAVELAB: http://www-stat.stanford.edu/~wavelab/

Appendix I
Other Versions of the HTK MFCC Filter-Bank

Table A.1 The narrowband version [0, 4000] Hz of the HTK filter-bank used in the HTK MFCC-FB20, reconstructed according to the description in (Young 1996)

Filter no.	Lower frequency [Hz]	Higher frequency [Hz]	Center frequency [Hz]	Filter bandwidth [Hz]
1	0	139	66	70
2	66	219	139	77
3	139	306	219	84
4	219	402	306	92
5	306	506	402	100
6	402	621	506	110
7	506	746	621	120
8	621	883	746	131
9	746	1033	883	144
10	883	1198	1033	158
11	1033	1378	1198	173
12	1198	1575	1378	189
13	1378	1791	1575	207
14	1575	2028	1791	227
15	1791	2287	2028	248
16	2028	2570	2287	271
17	2287	2881	2570	297
18	2570	3220	2881	325
19	2881	3593	3220	356
20	3220	4000	3593	390

T. Ganchev, *Contemporary Methods for Speech Parameterization*, SpringerBriefs
in Electrical and Computer Engineering, DOI 10.1007/978-1-4419-8447-0,
© Springer Science+Business Media, LLC 2011

Table A.2 The filter-bank used in the MFCC-FB26 as in the recent HTK implementations (Young et al. 2006)

Filter no.	Lower frequency [Hz]	Higher frequency [Hz]	Center frequency [Hz]	Filter bandwidth [Hz]
1	0	144	68	72
2	68	226	144	79
3	144	317	226	87
4	226	416	317	95
5	317	525	416	104
6	416	645	525	115
7	525	777	645	126
8	645	921	777	138
9	777	1080	921	152
10	921	1254	1080	167
11	1080	1445	1254	183
12	1254	1655	1445	201
13	1445	1886	1655	221
14	1655	2139	1886	242
15	1886	2416	2139	265
16	2139	2721	2416	291
17	2416	3056	2721	320
18	2721	3423	3056	351
19	3056	3827	3423	386
20	3423	4270	3827	424
21	3827	4756	4270	465
22	4270	5289	4756	510
23	4756	5875	5289	560
24	5289	6519	5875	615
25	5875	7225	6519	675
26	6519	8000	7225	741

References

Allen JB (1996) Harvey Fletcher's role in the creation of the communication acoustics. *Journal of the Acoustical Society of America* 99(4):1825–1839

Assaleh KT, Mammone RJ (1994a) Robust cepstral features for speaker identification. *Proceedings of the IEEE International Conference on Acoustics, Speech, and Signal Processing (ICASSP'94)*, Adelaide, Australia. Vol. 1, pp. 129–132

Assaleh KT, Mammone RJ (1994b) New LP-derived features for speaker identification. *IEEE Transactions on Speech and Audio Processing* 2(4):630–638

Atal BS, Hanauer SL (1971) Speech analysis and synthesis by linear prediction of the speech wave. *Journal of the Acoustical Society of America* 50(2):637–655

Atal BS (1974) Effectiveness of linear prediction characteristics of the speech wave for automatic speaker identification and verification. *Journal of the Acoustical Society of America* 55(6):1304–1312

Athineos M, Hermansky H, Ellis DPW (2004) LP-TRAP: Linear Predictive Temporal Patterns. *Proceedings of the ICSLP-2004*, Korea, Oct. 2004

Batliner A, Steidl S, Schuller B, Seppi D, Vogt T, Wagner J, Devillers L, Vidrascu L, Aharonson V, Kessous L, Amir N (2011) Whodunnit - Searching for the Most Important Feature Types Signaling Emotion-Related User States in Speech. *Computer Speech and Language* 25(1):4–28

Benesty J, Sondhi MM, Huang Y (Eds.) (2008) Springer Handbook of Speech Processing. ISBN: 978-3-540-49125-5, Springer-Verlag, Berlin, Heidelberg

Beranek LL (1949) Acoustic Measurements, New York, Wiley

Bogert BP, Hearly MJR, Tukey JW (1963) Quefrency analysis of time series for echoes: cepstrum, pseudo-autocovariance, cross-cepstrum and saphe cracking. *Proceedings of the Symposium on Time Series Analysis*, M. Rosenblatt, Rd. (John Wiley & Sons, Inc., New York, 1963), Chapter 15, pp. 209–243

Bogert BP (1967) Informal comments on the uses of power spectrum analysis. *IEEE Trans. on Audio and Electroacoustics* 15(2):74–75, June 1967

Bridle JS, Brown MD (1974) An Experimental automatic word-recognition system: Interim report. JSRU Report, No.1003, Dec. 1974

Campbell JP (1997) Speaker Recognition: A tutorial. *Proceedings of the IEEE*, 85(9):1437–1462, Sept. 1997

Campbell W, Sturim D, Reynolds D (2006) Support vector machines using GMM supervectors for speaker verification. *IEEE Signal Processing Letters* 13(5):308–311

Chen K, Wang L, Chi H (1997) Methods of combining multiple classifiers with different features and their applications to text-independent speaker recognition. *International Journal on Pattern Recognition and Artificial Intelligence* 11(3):417–445, 1997

Chistovich LA (1985) Central auditory processing of peripheral vowel spectra, *Journal of the Acoustical Society of America* 77:789–805, October 1985

Crandall IB (1917) The composition of speech. *Phys. Rev.* 10(1):74–76, July 1917

Crandall IB (1925) The sounds of speech. Bell System Technical Journal 4(4):586–626, Oct. 1925

Daubechies I (1992) Ten Lectures on Wavelets. SIAM, Philadelphia, PA, USA

Davis SB, Mermelstein P (1980) Comparison of parametric representations for monosyllabic word recognition in continuously spoken sentences. *IEEE Trans. on Acoustic, Speech and Signal Processing* 28(4):357–366

Deller JR, Proakis JG, Hansen JHL (1993) Discrete-time processing of speech signals. Prentice Hall, 1993

Dudley H (1939) Remaking speech. *Journal of the Acoustical Society of America* 11:169–177, October 1939

Erzin E, Cetin AE, Yardimci Y (1995) Subband analysis for speech recognition in the presence of car noise. *Proceedings of the IEEE International Conference on Acoustics, Speech, and Signal Processing (ICASSP-95)*, Detroit, MI, USA. Vol. 1, pp. 417–420

Eyben F, Woellmer M, Schuller B (2010) openSMILE: the Munich versatile and fast open-source audio feature extractor. *Proceedings of the International Conference on Multimedia*, ACM, New York, NY, USA, pp. 1459–1462

Fant G (1949) Analys av de svenska konsonantljuden. L.M. Ericsson protokoll H/P 1064 (139 pages)

Fant CGM (1956) On the predictability of formant levels and spectrum envelopes from formant frequences. In M. Halle, H. MacLean (Eds), *For Roman Jakobson*, Mouton & Co, The Hague, pp. 109–120

Fant G (1973) Speech sounds and features. The MIT Press. Cambridge, MA, USA

Farooq O, Datta S (2001) Mel filter-like admissible wavelet packet structure for speech recognition. *IEEE Signal Processing Letters* 8(7):196–198

Farooq O, Datta S (2002) Mel-scaled wavelet filter based features for noisy unvoiced phoneme recognition. *Proceedings of the 7th International Conference on Spoken Language Processing (ICSLP 2002)*, Denver, Colorado, USA. pp. 1017–1020

Fauve BGB, Matrouf D, Scheffer N, Bonastre J-F, Mason JSD (2007) State-of-the-art performance in text-independent speaker verification through open-source software. *IEEE Trans. on Audio, Speech, and Language Processing* 15(7):1960–1968

Flanagan JL (1972) Speech analysis, synthesis and perception. Springer-Verlag, Berlin

Fletcher H, Munson W (1933) Loudness, its definition, measurement, and calculation. *Journal of the Acoustical Society of America* 5:82–108

Fletcher H (1938a) Loudness, masking and their relation to the hearing process and the problem of noise measurement. *Journal of the Acoustical Society of America* 9:275–293

Fletcher H (1938b) The mechanism of hearing as revealed through experiment on the masking effect of thermal noise", *Proceedings of the National Academy of Sciences of the United States of America*, vol. 24, no. 7, Jul. 15, 1938, pp. 265–274. Available at: http://www.jstor.org/stable/87435

Fletcher H (1940) Auditory patterns. *Reviews of Modern Physics* 12:47–65, Jan. 1940. DOI: 10.1103/RevModPhys.12.47

Gabor D (1946) Theory of communication. *Journal of Institution of Electrical Engineers*, 93(3):429–457, November 1946

Ganchev T, Fakotakis N, Kokkinakis G (2002a) Text-independent speaker verification based on probabilistic neural networks. *Proceedings of the Acoustics 2002*, Patras, Greece. pp. 159–166

Ganchev T, Fakotakis N, Kokkinakis G (2002b) A speaker verification system based on probabilistic neural networks. *2002 NIST Speaker Recognition Evaluation, Results CD Workshop Presentations & Final Release of Results*, Vienna, Virginia, USA

Ganchev T (2005) Speaker Recognition. PhD dissertation. Dept. of Electrical and Computer Engineering, University of Patras, Greece, November 2005

Ganchev T, Fakotakis N, Kokkinakis G (2005) Comparative evaluation of various MFCC implementations on the speaker verification task. *Proceedings of the 10th International Conference on Speech and Computer, (SPECOM 2005)*, October 17–19, 2005. Patras, Greece, vol. 1, pp. 191–194.

Ganchev T, Mporas I, Fakotakis N (2010) Automatic height estimation from speech in real-world setup. *Proceedings of the 2010 European Signal Processing Conference (EUSIPCO 2010)*, Aalborg, Danmark, August 23–27, 2010, pp. 800–804

Garofolo J (1998) Getting started with the DARPA-TIMIT CD-ROM: An acoustic phonetic continuous speech database. National Institute of Standards and Technology (NIST), Gaithersburgh, MD, USA

Glasberg BR, Moore BCJ (1990) Derivation of auditory filter shapes from notched-noise data. *Hearing Research* 47(1–2):103–138

Greenberg G, Martin A, Brandschain L, Campbell J, Cieri C, Doddington G, Godfrey J (2010) Human assisted speaker recognition (HASR) in NIST SRE 2010. *Proceedings of the speaker and language recognition workshop, Odyssey 2010*, June 28-July 1, 2010, Brno, Czech Republic

Greenwood DD (1991) Critical bandwidth and consonance in relation to cochlear frequency –position coordinates. *Hearing research* 54:165–208

Greenwood DD (1997) The Mel Scale's disqualifying bias and a consistency of pitch-difference equisections in 1956 with equal cochlear distances and equal frequency ratios. *Hearing research* 103:199–248

Grezl F, Karafiat M, Cernocky J (2004) TRAP based features for LVCSR of meeting data. *Proc. of ICSLP-2004*, Korea, Oct. 2004

Harris FJ (1978) On the use of windows for harmonic analysis with the discrete Fourier transform. *Proceedings of the IEEE* 66(1):51–83, January 1978

Hennebert J, Melin H, Petrovska D, Genoud D (2000) POLYCOST: A telephone-speech database for speaker recognition. *Speech Communication* 31(2–3):265–270

Hermansky H (1990) Perceptual linear predictive (PLP) analysis for speech. *Journal of the Acoustical Society of America* 87(4):1738–1752

Hermansky H (2003) TRAP-TANDEM: Data-driven extraction of temporal features from speech. Technical Report IDIAP-RR-03-50, August 31, 2003

Huang X, Acero A, Hon HW (2001) Spoken language processing: A guide to theory, algorithm, and system development. Prentice Hall

Kenny P, Ouellet P, Dehak N, Gupta V, Dumouchel P (2008) A study of inter-speaker variability in speaker verification. *IEEE Trans. Audio Speech and Language Processing* 16(5):980–988

Kim DS, Lee SY, Kil RM (1999) Auditory processing of speech signals for robust speech recognition in real-world noisy environments. *IEEE Trans. Speech Audio Processing* 7(1):55–69

King IR (1971) A comparison of existing eigenvector studies of the dimensionality of speech. Institute for Defense Analyses, Princeton, N.J., Communication Research Division, Working Paper No. 333, Sept. 1971.

Kinnunen T, Li H (2010) An overview of text-independent speaker recognition: from features to supervectors. *Speech Communication* 52(1):12–40

Klein W, Plomp R, Pols LC (1970) Vowel spectra, vowel spaces, and vowel identification. *Journal of the Acoustical Society of America* 48(4):999–1009

Koenig W (1949) A new frequency scale for acoustic measurements. *Bell Telephone Laboratory Record* 27:299–301

Lee K-F, Hon H-W, Reddy R (1990) An overview of the SPHINX speech recognition system. *IEEE Trans. Acoustics Speech and Signal Processing* 38(1):35–45

LePage EL (2003) The mammalian cochlear map is optimally warped. *Journal of the Acoustical Society of America* 114(2):896–906, August 2003

Long CJ, Datta S (1996) Wavelet based feature extraction for phoneme recognition. *Proceedings of the ICSLP-96*, Philadelphia, USA. Vol. 1, pp. 264–267

Mattingly IG (1999) A short-history of Acoustic Phonetics in the U.S. *Proceedings of the 14th International Congress of Phonetic Sciences*, San Francisco, CA, USA, pp. 1–6

Makhoul J (1975) Spectral linear prediction: properties and applications. *IEEE Trans. on Acoustics Speech and Signal Processing* 23:283–296

Mermelstein P (1976) Distance measures for speech recognition, psychological and instrumental. In *Pattern Recognition and Artificial Intelligence*, C.H. Chen, Ed., New York, Academic Press, pp. 374–388

Miller DC (1916) Science of the musical sounds. Macmillan, New York

Miller JD (1989) Auditory-perceptual interpretation of the vowel. *Journal of the Acoustical Society of America* 85(5):2114–2134, May 1989

Moore BCJ, Glasberg BR (1983) Suggested formulae for calculating auditory-filter bandwidths and excitation patterns. *Journal of the Acoustical Society of America*, 74(3):750–753

Moore BCJ, Glasberg BR (1996) A revision of the Zwicker's loudness model. *Acustica-Acta Acustica*, 82:335–345

Moore BCJ (2003) An introduction to the psychology of hearing. Academic Press, London, 5th Ed.

Mporas I, Ganchev T, Siafarikas M, Fakotakis N (2007) Comparison of speech features on the speech recognition task. *Journal of Computer Science*, 3(8):608–616

Mporas I, Ganchev T, Fakotakis N (2008) Phonetic segmentation using multiple speech features. *International Journal of Speech Technology* 11(2):73–85, June 2008

Mporas I, Ganchev T (2009) Estimation of unknown speaker's height from speech. *International Journal of Speech Technology* 12(4), December 2009

Nogueira W, Büchner A, Lenarz T, Edler B (2005) A psychoacoustic "NofM"-type speech coding strategy for cochlear implants. *EURASIP Journal on Applied Signal Processing – Special Issue on DSP in Hearing Aids and Cochlear Implants* 18:3044–3059

Nogueira W, Giese A, Edler B, Büchner A (2006) Wavelet packet filter-bank for speech processing strategies in cochlear implants. *Proceedings of the IEEE International Conference on Acoustics, Speech, and Signal Processing, (ICASSP 2006)*, Toulouse, France, vol. 5, pp. 121–124

NIST (2001) The NIST year 2001 speaker recognition evaluation plan. National Institute of Standards and Technology of USA. Available: http://www.nist.gov/speech/tests/spk/2001/doc/2001-spkrec-evalplan-v05.9.pdf

NIST (2002) The NIST year 2002 speaker recognition evaluation plan. National Institute of Standards and Technology of USA. Available: http://www.nist.gov/speech/tests/spk/2002/doc/2002-spkrec-evalplan-v60.pdf

Noll AM (1964) Short-time spectrum and "cepstrum" techniques for vocal-pitch detection. *Journal of the Acoustical Society of America* 36(2):296–302, Feb. 1964

Noll AM, Schroeder MR (1964) Short-time "cepstrum" pitch detection. *Sixty-Seventh Meeting of the Acoustical Society of America*, May 6–9, 1964, *Journal of the Acoustical Society of America* 36(5):1030, May 1964 (abstract)

Noll AM (1967) Cepstrum pitch determination. *Journal of the Acoustical Society of America* 41(2):293–309, Feb. 1967

Nuttall AH (1981) Some windows with very good sidelobe behavior. *IEEE Trans. on Acoustics, Speech, and Signal Processing* 29(1):84–91, Feb. 1981

Oppenheim AV (1967) Deconvolution of Speech, *Journal of the Acoustical Society of America* 41:1595, 1967 (abstract)

Oppenheim AV, Schafer RW, Stockham TG (1968) Non-linear filtering of multiplied and convolved signals. *Proceedings of the IEEE* 56(8):1264–1291, Aug. 1968

Oppenheim AV, Schafer RW (1968a) Homomorphic analysis of speech. *IEEE Trans. Audio and Electroacoustics*, 16(2):221–226, June 1968

Oppenheim AV, Schafer RW (1968b) Non-linear filtering of multiplied and convolved signals. *IEEE Trans. Audio and Electroacoustics*, 16(3):437–446, Sept. 1968

Oppenheim AV (1969) Speech analysis-synthesis system based on homomorphic filtering. *Journal of the Acoustical Society of America* 45(2):458–465

Oppenheim AV, Schafer RW (2004) From frequency to quefrency: A history of the cepstrum. *IEEE Signal Processing Magazine* 21(5):95–99 & 106.

O'Shaughnessy D (1987) Speech communications: Human and machine. Addison-Wesley Publishing Co., Reading, MA, USA

Patterson RD, Moore BCJ (1986) Auditory filters and excitation patterns as representation of frequency resolution. In B.C.J. Moore (Ed.) Frequency Selectivity in Hearing, Academic Press. London, pp. 123–177

Pols LCW (1971) Real-time recognition of spoken words. *IEEE Trans. on Computers*, 20(9):972–978, Sept. 1971

Reynolds DA (1994) Experimental evaluation of features for robust speaker identification. *IEEE Trans. on Speech and Audio Processing* 2(4):639–643, Oct. 1994

Sarikaya R, Gowdy JN (1998) Subband based classification of speech under stress. *Proceedings of the Int. Conf. Acoustics, Speech, and Signal Processing, ICASSP'98*, vol. 1, Seattle, WA, 1998, pp. 569–572

Sarikaya R, Pellom BL, Hansen JHL (1998) Wavelet packet transform features with application to speaker identification. *Proceedings of the IEEE Nordic Signal Processing Symposium*, Vigso, Denmark, June 1998, pp. 81–84

Sarikaya R, Hansen JHL (2000) High resolution speech feature parameterization for monophone-based stressed speech recognition. *IEEE Signal Proc. Let.* 7(7):182–185

Siafarikas M, Ganchev T, Fakotakis N (2004) Wavelet packets based speaker verification. *Proc. of the Odyssey 2004*, Toledo, Spain. pp. 257–264

Siafarikas M, Ganchev T, Fakotakis N, Kokkinakis G (2005) Overlapping wavelet packet features for speaker verification. *Proceedings of the INTERSPEECH'05*, Lisbon, Portugal, pp. 3121–3124

Siafarikas M, Ganchev T, Fakotakis N, Kokkinakis G (2007) Wavelet packet approximation of critical bands for speaker verification, *International Journal of Speech Technology* 10 (4):197–218, 2007

Slaney M (1998) Auditory toolbox. Version 2. Technical Report #1998-010, Interval Research Corporation

Skowronski MD (2004) Biologically inspired noise-robust speech recognition for both man and machine. Ph.D. Dissertation, University of Florida, 2004

Skowronski MD, Harris JG (2004) Exploiting independent filter bandwidth of human factor cepstral coefficients in automatic speech recognition. *Journal of the Acoustical Society of America* 116(3):1774–1780

Schroeder MR (1977) Recognition of complex acoustic signals. *Life Science Research Reports*, T.H. Bullock, Ed., 55:323–328

Schroeder MR, Atal BS, Hall JL (1979) Optimizing digital speech coders by exploiting masking properties of the human ear. *Journal of the Acoustical Society of America* 66(6):1647–1652

Schuller B, Steidl S, Batliner A (2009) The INTERSPEECH 2009 emotion challenge. *Proc. of Interspeech 2009*, Brighton, UK, pp. 312–315

Schuller B, Steidl S, Batliner A, Burkhardt F, Devillers L, Muller C, Narayanan S (2010) The INTERSPEECH 2010 paralinguistic challenge - age, gender, and affect. *Proceedings of the 11th International Conference on Spoken Language Processing, INTERSPEECH 2010 - ICSLP*, Makuhari, Japan, 2010, pp. 2794–2797

Stevens SS, Volkman J, Newman EB (1937) A scale for the measurement of the psychological magnitude pitch. *Journal of the Acoustical Society of America* 8(1):185–190, Jan. 1937

Stevens SS, Volkman J (1940) The relation of pitch to frequency: A revised scale. *American Journal of Psychology* 53(3):329–353, July 1940

Stevens SS (1957) On the psychophysical law. *Psychology Review* 64:153–181

Tufekci Z, Gowdy JN (2000) Feature extraction using discrete wavelet transform for speech recognition. *Proceedings of the IEEE SoutheastCon 2000*, Nashville, Tennessee, USA. pp. 116–123

Umesh S, Cohen L, Nelson D (1999) Fitting the Mel scale. *Proceedings of the ICASSP-99*, Phoenix, USA, 15–19, March, vol. 1, pp. 217–220

Valente F (2010) Hierarchical and parallel processing of auditory and modulation frequencies for automatic speech recognition. *Speech Communication* 52:790–800

Vertanen K (2006) Baseline WSJ acoustic models for HTK and Sphinx: Training recipes and recognition experiments. Technical Report, Cavendish Laboratory, 2006

Yilmaz H (1967) A theory of speech perception. *Bulletin of Mathematical Biophysics* 29(4):739–825

Young SJ, Odell J, Ollason D, Woodland P (1995) The HTK Book. Version 2.0. Department of Engineering, Cambridge University, UK

Young SJ (1996) A review of large-vocabulary continuous-speech recognition. *IEEE Signal Processing Magazine* 13(5):45–57, September 1996

Young SJ, Evermann G, Gales M, Hain T, Kershaw D, Liu X, Moore G, Odell J, Ollason D, Povey D, Valtchev V, Woodland P (2006) The HTK Book. Version 3.4. Department of Engineering, Cambridge University, UK

Zheng F, Zhang G, Song Z (2001) Comparison of different implementations of MFCC. *Journal of Computer Science and Technology* 16(6):582–589, Sept. 2001

Zwicker E, Flottorp G, Stevens SS (1957) Critical bandwidth in loudness summation. *Journal of the Acoustical Society of America* 29:548–557

Zwicker E (1961) Subdivision of the audible frequency range into critical bands (Frequenzgruppen). *Journal of the Acoustical Society of America* 33:248–249

Zwicker E, Terhardt E (1980) Analytical expressions for critical-band rate and critical bandwidth as a function of frequency. *Journal of the Acoustical Society of America* 68(5):1523–1525

Zwicker E (1981) This week's citation classics [Zwicker E, Flottorp G, Stevens SS (1957) Critical bandwidth in loudness summation. *Journal of the Acoustical Society of America* 29:548–557]", *Citation classic* 9, March 2, 1981